Carolina Journeys

Exploring the trails of the Carolinas—both real and imagined

by Tom Fowler

Copyright © 2004 by Gail Fowler
photos by Tom Fowler
All Rights Reserved

available from:
Parkway Publishers, Inc.
P. O. Box 3678
Boone, NC 28607
www.parkwaypublishers.com
Tel/Fax: 828-265-3993

Library of Congress Cataloging-in-Publication Data:

Fowler, Tom, 1955-
Carolina journeys : exploring the trails of the Carolinas—both real and imagined / by Tom Fowler.
  p. cm.
ISBN 1-887905-86-3
1. North Carolina—History, Local. 2. South Carolina—History, Local. 3. Roads—North Carolina. 4. Roads—South Carolina. 5. Historic sites—North Carolina. 6. Historic sites—South Carolina. 7. North Carolina—Description and travel. 8. South Carolina—Description and travel. I. Title.

F255.F69 2004
975.6—dc22
    2004005510

Cover Design by Aaron Burleson
Editing, layout and book design by Julie Shissler

Introduction . . . . *xi*

Union Square. *1*
How North Carolina's capital came to be located
on the high ground at Wake Crossroads, or, pour me another
one of them cherry bounces, Joel.

If It Hadn'a Been For Grayson. *6*
Tom Dooley's last road trip as a free man and his brief encounter
with Colonel Grayson.

*Retracing the Great Indian Trading Path:
Occaneechi Town to the Trading Ford* *12*
Searching for the path John Lawson followed in 1701
as he traveled across the Piedmont.

Live Oaks with First Names *18*
A few grand old oak trees still preside over the Carolina coast.

Sassafras Mountain *21*
A family hikes to the summit of the highest mountain in
South Carolina—in pursuit of a larger goal.

Monarchs of the Blue Ridge *25*
By November of each year all the monarch butterflies have fled
their summer home in the Carolinas—where do they go?

The Road to Adshusheer *31*
Walking in the 300 year old footsteps of John Lawson and Eno
Will—a reenactment hike from Hillsborough to West Durham.

The Shut-In Trail *40*
Running through rhododendron thickets on
Vanderbilt's Shut-In Trail.

History on a Stick 46
Reflections on a few of North Carolina's 1,434 historical markers.

Wolfe's Angels. 50
A marble angel in a Hendersonville cemetery may have inspired
the title of Thomas Wolfe's novel Look Homeward, Angel.

A Cary Hash: Tiny Tank's Tick Tour 53
Following a flour-marked trail through the new subdivisions of
Cary with a group of social runners and descriptions of the
resulting encounters with certain displaced and surly natives.

Encounters with the Shad Fish and the Shad Bones. . . 56
Getting intimate with a famous and historic
spring-time river runner.

Barbecue Church, Cross Creek, and Flora Mac 61
A world famous Scottish Highlander's turbulent
five year sojourn in North Carolina.

Into the Waccamaw 68
Disturbing the Waccamaw silverside and killifish
in a mysterious Carolina Bay.

Running the Scenic Crest of the Sauratown Mountains 74
A trail run from the summit of Pilot Mountain
to the summit of Hanging Rock.

Burnt Churches in South Carolina's Lowcountry 78
The peaceful ruins of several eighteenth century churches
stubbornly stand still.

Stalking the Elusive Carolina Petroglyph 81
The spirited pursuit of a remarkably evasive prey.

Tarleton's Tea Table *87*
    Al fresco dining in Lincoln County with Lord Cornwallis
    and Colonel Tarleton in the winter of 1781.

The View from the Firetower on Flat Top Mountain . . *90*
    Carriageless rambles up the gentle carriage trails of
    Moses and Bertha Cone to a tower on the summit
    of a mountain high above Blowing Rock.

Peter Ney *93*
    People who knew him routinely described him as
    the greatest man they'd ever met. What was it with this
    Rowan County schoolteacher who died in 1846?

The Night of the Shooting Stars *97*
    Neutrinos, quarks, James Joyce and celebrating the
    Leonid Meteor Shower with a few of our closest friends.

The Forks of the Yadkin *103*
    Before Kentucky, Daniel Boone roamed the woods
    of North Carolina.

After the Olympics *107*
    How Mia Hamm and her crowd have changed
    the American runner's world.

Judaculla Rock *109*
    The petroglyph hunter visits a big and significant rock
    in western North Carolina, where a great seven-toed slant-eyed
    giant once prowled.

The Bunker Brothers of Surry County *113*
    World famous twins seek the life of gentlemen farmers
    in North Carolina.

The Silver Cup *116*
A recollection of events leading up to, and some events subsequent to, a 1964 sailboat race on Kerr Lake — and the memorable prize awarded to the first place finisher.

Hellespont Dreams: Cross-Training with Lord Byron . *121*
A runner paddles into history by recreating Byron's famous swim across the Hellespont in 1810.

Faith Rock *125*
A visit to a famous rock that played a crucial role in a patriot's escape from the clutches of the notorious Revolutionary War loyalist, David Fanning, in 1781.

Pausing Beside the King's Highway *129*
A visit to a famous oak tree in Pender County.

The Bigfoot Chapel Hill Hash *132*
Following a flour-marked trail through a quaint college town with the Tar Heel Hash House Harrier association of social runners and imbibers.

Going After Elisha *134*
Dr. Elisha Mitchell's search for the highest mountain in the East, and his final hike to Mitchell Falls.

The Dismal Swamp Canal *140*
A remarkable ditch connecting Chesapeake Bay with Albemarle Sound.

De Soto Slept Here *144*
In the 1500's several expeditions of Spanish soldiers traveled up from Florida into parts of North Carolina— but their precise paths are still debated.

Meadowmont, Turtle-Handling & the Havacow Hash 150
Encounters with a large turtle and the future Meadowmont of Chapel Hill while following a flour marked trail with the Tar Heel Hash House Harrier association of social runners and imbibers.

Introduction to Carolina Journeys

In the mid-1980s while traveling in China, I visited a section of the Great Wall that the Chinese authorities were transforming into a major tourist attraction. The chairlift from the parking lot up the mountain was still under construction, but a dusty trail up the side of the ridge got us to the Wall. The section of the Wall we reached was in remarkable shape for such an ancient structure. The bricks were all in place, the grout smooth and even, and the guard towers swept clean of dust and debris. The Wall had clearly been restored. Even more impressive, however, was how the Wall kept going, snaking up the incredibly steep slopes on either side of the low point of the ridge on which we stood. And beyond the nearest summits, the Wall continued up narrow ridges and cresting the sharp mountains as far as we could see. Quite remarkable. I was drawn to climb farther along the Wall as it headed up the mountain.

The renovated section of the Wall soon ended and I was hiking on an unreconstructed Great Wall. Small trees and bushes covered the top of the Wall with a narrow dirt path winding upward amid piles of brick and stone. The watch towers were crumbling and remains of stone arched doorways stood out against the sky. This was the real ruin of the Wall—what was left of this magnificent construction after hundreds of years (although the Wall has existed for over 2,000 years, it owes its present form to major renovations made about 500 years ago). There was no telling what I was going to see before I saw it or sometimes what it was even after I saw it. The climbing was steep, but I kept on. Finally at a parapet on the Wall I turned to look back. The view was spectacular. The reconstructed part of the Wall was far below and I could see even more of the Wall coiling around the mountains off into the distance. A faint shout carried up to me. Most of my tour group was still down on the lowest part of the Wall. Some were waving. Without making out their words, I understood. My hike was over. Regretfully I began my descent.

Now the restored Great Wall was impressive and I would not have missed it for the world. But the unreconstructed Wall was riveting. Time had slowed to a crawl as I trudged up the old Wall looking at everything so closely. Was it because I didn't know what I was supposed to see? Did my lack of expectation and of preconception, by making everything potentially significant, focus my mind entirely on the present? The rebuilt Wall had been sanitized, standardized, evaluated and interpreted by authorities and experts. It required little of me except to view it and then to compare it to what I'd been told I should see. The old Wall, however, required my attention, my own analysis of what I saw, and my own conclusions. Small wonder that those who never left the smooth, rebuilt brick walk of the restored Wall, having seen the sight, found their thoughts turning toward the bus ride back to Beijing—and how soon lunch would be.

The novelist Walker Percy thought about this effect of expectation upon perception. He described it as surrendering sovereignty over the experience to the expert—or at least to someone else who will evaluate your experience for you even before you experience it. In *The Message in the Bottle*, Percy considered a man from Boston who decides to vacation at the Grand Canyon. He visits his travel agent, looks at all the pamphlets and signs up for a tour of the Canyon. Percy believed this man would find it impossible upon arriving at the Bright Angel Lodge, to gaze directly at the Grand Canyon and see it for what it is. Impossible because, for this man, the Canyon would have been

> appropriated by the symbolic complex which has already been formed in the sightseer's mind. Seeing the canyon under approved circumstances is seeing the symbolic complex head on. The thing is no longer the thing as ... confronted ...; it is rather that which has already been formulated—by picture postcards, geographic book, tourist folders, and the words Grand Canyon. ... The highest point, the term of the sightseer's satisfaction, is not the sovereign discovery of the thing before him; it is rather the measuring up of the thing to the criterion of the preformed symbolic complex.

Percy's thesis may help explain why for some of us there is something dissatisfying about visiting the well-known tourist destinations in the Carolinas. These sites may be too well interpreted and packaged for us by those who know better than us. All we need do, indeed all we are implicitly allowed to do, is to follow the excellent directions to the site, view the site by following the walking tour, and compare what we've seen to what we were expecting to see. There is nothing to be discovered, nothing to be explored, nothing to be figured out. It's all been done for us.

Luckily, however, the Carolinas are also full of poorly-known tourist destinations that have not been well interpreted or well packaged for us by those who know better. These sites may be hard to find and information about the sites may be hard to come by—sometimes requiring additional effort and often causing additional aggravation. But the trade-off is that visitors to these sites may regain sovereignty over their experience. They may regain that sense of exploration and discovery that can expand and enrich perception, and they may regain the excitement of searching for something you aren't sure you will find.

There are plenty of publications that provide information about the well-known tourist destinations in the Carolinas. Read them at your peril. They will co-opt your sovereignty and reduce you to a sightseer. *Carolina Journeys*, on the other hand, is intended to tell stories of poorly-known sites of interest in the Carolinas. We realize that providing information about these sites is the first step in co-opting your sovereignty and reducing you to a sightseer—so our goal is to avoid providing too much information or too good directions or being too knowledgeable and authoritative. Much will be left up to you, dear reader and Carolina sojourner. It is, of course, for your own good—not because we are lazy. This is, after all, the whole point of this introduction. *Carolina Journeys* will speak of history, culture and specific places—but we also hope our stories will help inspire our readers to perform their own weekend explorations to discover or rediscover the different trails, real and imagined, that can be found throughout the Carolinas. And we hope you will share with us a little (but not too much) about your travels.

Walker Percy observed: "Every explorer names his island Formosa, beautiful. To him it is beautiful because, being first, he has access to it and can see it for what it is. But to no one else is it ever as beautiful—except the rare man who manages to recover it, who knows that it has to be recovered."

Tom Fowler

Little in the Carolinas remains to be discovered, but much continues to be quite beautiful—if we can regain the ability to see it for what it is. *Carolina Journeys* is in the recovery business.

Tom Fowler
Durham, North Carolina
www.carolinajourneys.com

The unreconstructed Great Wall of China

Union Square

For several years now I've enjoyed a fourth floor office in downtown Raleigh with a grand view of the North Carolina State Capital just across Morgan Street from my building. When I look up from my computer to stretch my neck and ponder some difficult aspect of my work for the state, I often glance out my window and watch the flags flapping on the Capital dome or the occasional demonstrators carrying their signs and bullhorns around the statues on the Capital grounds. Sometimes I glimpse the brown and white hawk that sits high up in the big trees surrounding the Capital eyeing the busy but scatter-brained resident squirrels. And then sometimes, as my mind relaxes and wonders, I gaze out the window past the Capital building just beyond the park-like Capital grounds to where the brown waters of the stately Neuse River flow by.

And sometimes in early spring I watch the shad fishermen on the banks of the Neuse, in the shadow of the state legislative building, pulling in scads of the bony fish and tossing them in wicker baskets. And then the telephone rings, I snap back to reality and my vision of the Neuse loses focus and then evaporates. I recall that shad can't swim this far upstream because of all the dams across the Neuse blocking their way. And in any event, I have to admit that despite how much I wish it weren't so, the commissioners meeting at Joel Lane's house in 1792 did indeed change their minds at the last minute and, instead of selecting for the location of the state capital the lovely once-favored site on the banks of the Neuse, they voted for the riverless Wake Crossroads site where the Capital Building now stands in the center of Union Square in the heart of Raleigh. And fifteen miles away to the east, old man Neuse still flows, ponderous and majestic, on its way to the sea through the empty wooded countryside that might once have been Raleigh.

In its early years the North Carolina legislature would convene all over the eastern half of the state. It met at least once in Bath, Brunswick, Halifax, New Bern, Smithfield, Tarboro and Wilmington. But in 1788 the legislature met in Hillsborough and decided that North Carolina would have an "unalterable seat of government" and that it should be located "within 10 miles of the plantation whereon Isaac

Hunter now resides"—Hunter owned a popular tavern in Wake County. In 1792, the legislature appointed nine commissioners to meet in Wake County and hammer out the precise size and location of the future capital. Some of the commissioners' names might ring a bell with those familiar with the street names in downtown Raleigh, e.g., Blount, Dawson, Hargett, Jones, Martin, McDowell, Person.

In late March of 1792 the commissioners journeyed to Wake County to begin their work. They gathered at the home of Joel Lane, by all accounts the grandest house in the sparsely populated Wake County. Lane called his home "Wakefield." The commissioners would stay at Wakefield for the next eight days, apparently traveling around the county examining various parcels of land during the days and retiring to Lane's house at night to discuss what they'd seen and to recover their energy and spirit courtesy of Lane's famous hospitality.

Today, if you drive past North Carolina State University toward downtown Raleigh on Hillsborough Road, after a couple of stoplights you'll see St. Mary's School and then St. Mary's Road. If you turn right on St. Mary's Road, in two short blocks you'll dead-end onto West Hargett Street. And there on your right is Wakefield—the house Joel Lane built in 1760 and the very house where the commissioners met in 1792. It's the same house but in 1917 it was moved a short distance from its original location. The original site of Wakefield is located about a block or so to the northeast of the house's present location. The knoll on which the house was built is still discernible where Morgan Street dips downhill before its intersection with Boylan Avenue. In the late 1700s Lane owned several thousands of acres in Wake County—but he built his house to be close to the intersection of two major highways—one running north to south and connecting Petersburg and Fayetteville, and the other running east to west and connecting New Bern and Hillsborough. The highways intersected about a half mile east of Lane's house, at a spot known as Wake Crossroads. It's said that in 1792 travelers and drovers used Wake Crossroads as a place to camp and rest their herds of livestock during their dusty trips along these two great roads. Joel Lane owned Wake Crossroads and all the land surrounding it.

History tells us that the early votes of the commissioners during their sojourn at Wakefield favored locating the permanent capital on the north bank of the Neuse River on land owned by Col. John Hinton—a relative by marriage of Joel Lane. Some retellings of the events claim that even on the last night before the final decision, the Neuse site was still the favorite. But on that last night, Joel Lane served his guests his special, thirst-quenching drink called a "cherry bounce"—maybe some sort of brandy or cherry wine—and surely there was much grand conversation and good cheer at Wakefield that night. In the morning the commissioners made their final vote and decided to fix the North Carolina capital on the land owned by their host, Joel Lane, with its center on the campground known as Wake Crossroads. The commissioners agreed to buy 1,000 acres of Lane's land for 1,378 pounds—approximately $2,756. Raleigh would be a riverless capital.

Lane deeded the property to the state on April 5, 1792. Surveyors laid out the new town of Raleigh into a rectangle of ten by eleven blocks with Wake Crossroads becoming Union Square just to the north of the rectangle's center. Union Square and the nearby lots were reserved for the state capital and the other lots were auctioned off to raise money for construction of the capital building. The first state capital building was completed in 1796 but was destroyed by fire in 1831. Construction of the Capital Building that now stands in the center of Union Square began in 1833 and was completed in 1840. Until 1888 the Capital Building housed all three branches of state government: the executive branch on the first floor; the legislative branch on the second floor; and the Supreme Court on the third floor.

Sometimes after work on a hot summer's day, I'll leave the back door of my building and walk past the statue of Sir Walter Raleigh that stands at the north end of Fayetteville Street Mall. Sir Walter's statue proudly looks north toward Union Square and the State Capital. Following his glance, I'll cross Morgan Street and stride up the steps onto the grounds of the old drovers' campsite, the former Wake Crossroads, now Union Square (and sometimes called Capital Square). The crossroads are still there. Fayetteville Street (now a pedestrian mall), the old highway to Fayetteville and Charleston, stretches southwards behind me. New Bern Avenue heads toward the coast from the eastern side of Union Square. Hillsborough Road begins in the middle of the western border of the Square, and to the north—

now a pedestrian mall leading to the state legislature—is the path to Halifax and Petersburg, Virginia, shown on old Raleigh maps as Halifax Road.

Often I'll circumambulate the capital grounds. I'll stop at the first statue placed on Union Square—a statue of George Washington, on the south side of the Capital Building, placed there in 1858 and flanked by two cannon that were mounted at Edenton in the 1770s. I'll pass by the standing stone U.S. Geodetic Survey markers (placed there in 1853) in the Square's southeastern corner close to the state's geographic center. On the eastern side of the Square I'll walk by the three presidents monument and the two mortars used at Fort Macon during the Civil War. And just to the northwest of the Capital Building, I'll walk by the statue of the on-rushing Henry Lawson Wyatt, the first North Carolina soldier killed in the Civil War—shot at the Battle of Bethel Church on June 10, 1861. Then I'll walk past the spot where I once watched the Union Square hawk standing in the grass next to the sidewalk calmly munching on a squirrel as state workers carefully walked by. I'll look up into the Square's big hardwoods to see if the brown and white bird is standing sentinel over the Square but usually I can't find him.

Then I'll leave the park-like grounds of Union Square, reach my parking place, and head out on the old Hillsborough Road toward my home to the west. I'll cross over West Street (the western boundary of the rectangle first surveyed in 1792) and then pass by the old site of Lane's Wakefield on the knoll overlooking downtown Raleigh. But I won't stop—even though I'm thirsty and I could use icy, cherry-flavored thirst-quenching refreshment. Maybe something like a drink called a cherry bounce—even if made with Wake County well water rather than river water from old man Neuse. But, I know, I'm a couple hundred years too late to place my order. So I'll just keep going past Hargett Street. And I'll pull in at the next Fast Fare and pick up a can of Cheerwine for the drive home.

Carolina Journeys

Statue of George Washington, Union Square

If It Hadn'a Been For Grayson

He met her on a mountain. And that's where he killed her. Did it with his knife, he did. Well now, boys. Maybe it's no surprise to you that the hang down your head miscreant made famous by the Kingston Trio's song was a real fellow who grew up in Wilkes County, North Carolina, and fought with the South in the Civil War. After the war Tom Dooley returned home to Wilkes County—but he had some trouble settling down. He had some problems—women problems. Tom Dooley was twice tried and twice convicted of murdering a local girl, Laura Foster, with his knife, and he was sentenced to death by hanging. But he wasn't executed down in some lonesome valley, swinging from a white oak tree, as the song suggests. He was hung from a scaffold in front of hundreds of spectators in downtown Statesville on May 1, 1868. But it is true that except for a guy named Grayson, Tom Dooley might have lived out his natural life in Tennessee.

Tom Dooley (actually spelled "Dula" though apparently pronounced "Dooley") was a handsome, banjo-pickin' (or maybe a fiddle-playin') n'er do well. After returning from the war, he lived with his mother and sister on Reedy Branch, close by the little community of Elkville which had grown up where Elk Creek flows into the Yadkin River in western Wilkes County. In 1866, he was a twenty-two year old with a complicated love life—probably more so than we will ever know. He had several simultaneous romances. These included Laura Foster, and the woman who may have been the real love of his life, Ann Melton—a beautiful, but married, young woman who lived nearby.

Early on the morning of May 25, 1866, Laura Foster packed some clothes, saddled her father's horse and left her home in Caldwell County and headed for Elkville. On the way Laura met an acquaintance and told her she was riding to meet Tom Dooley that morning at the Bates place near Elk Creek and that she and Tom planned to marry. Several people saw Tom Dooley in the vicinity of the Bates place on that same morning. Later that day Tom was seen near the Bates place with a digging implement known as a mattock. No one ever saw Laura alive again.

On the morning of 26 May, Laura's horse returned on its own to her home in Caldwell County. The horse's lead rope was broken and dangled from the halter. A search for Laura in the vicinity of the Bates place proved fruitless. In late June the other end of the lead rope was found tied to a dogwood tree near the Bates place. People began to say openly that Tom Dooley had murdered Laura. On the night of June 25, 1866, Tom said goodbye to a tearful Ann Melton and left Elkville. More on that later.

In late August of 1866, based on public comments she had made, Ann Melton's cousin, Pauline Foster, was arrested on suspicion of being an accessory to the murder of Laura Foster. Jail worked its magic on Pauline and in early September she led a search party to a shallow grave on a ridge above the Bates place. Two and a half feet down, they found Laura Foster's corpse together with the bundle of clothes she had taken with her on that late May morning. She still wore her checkered cotton dress. Laura had been stabbed in the left breast between the third and fourth ribs. Ann Melton was arrested soon afterward. Tom Dooley had already been chased down and caught in Tennessee on July 10. He was returned to the Wilkesboro jail a few days later where he was held without bail. After the body was found, Dooley was officially indicted for the murder, as was Ann Melton.

Justice was swifter in those days. Ann Melton's request to be tried separately was granted and Tom Dooley's trial began the next month, on October 19, 1866. He was represented by the former governor of North Carolina, Zebulon B. Vance. The trial was completed in three days and Dooley was convicted and sentenced on October 21. But Attorney Vance appealed on Dooley's behalf and the North Carolina Supreme Court ordered that Tom Dooley deserved a new trial. Dooley was tried again and convicted again in January of 1868. Vance appealed once again to the Supreme Court but this time the verdict was upheld. Tom Dooley would hang.

The evidence presented by the prosecution suggested that Tom Dooley and Ann Melton conspired to murder Laura Foster, perhaps because of a social disease that they believed each of them had contracted as a result of Tom's relationship with Laura. Only Ann and Tom knew, however, whether it was Tom or Ann who actually stabbed Laura. Pauline Foster's testimony, if true, did appear to establish that Ann Melton was at least an aider and abettor of the crime. Nevertheless, at her trial which took place after Tom's hanging, Ann Melton was found not guilty. There are those who still claim that the evidence against Tom was all circumstantial and that he was not guilty of the murder of Laura Foster. Seems unlikely to me though. But it's also unlikely that we will ever know what really happened on Laura Foster Ridge on May 25, 1866. We do know something, however, about Tom Dooley's last lonesome road trip—when the public accusations of his involvement with Laura's murder caused him to flee Elkville and head for Tennessee.

In the evening of June 25, 1866, Tom Dooley visited Ann Melton for the last time as a free man. Pauline Foster, who lived with the Melton's and was present during this visit, said Tom seemed depressed, and that Tom and Ann lay down on the bed together and both began crying. Pauline asked Tom what was wrong and he said they were telling lies about him and that he was going to have to leave the county. He said he would return by Christmas to take both his mother and Ann Melton away with him. Later that night Tom embraced the weeping Ann for the last time and he left Elkville, walking north toward Watauga County.

I'm driving southwest from Wilkesboro on State Road 268 which runs besides the Yadkin River. I drive across Beaver Creek. Daniel Boone once lived near here—where Beaver Creek flows into the Yadkin. Soon I see the historical marker for Thomas C. Dula, which states that he is buried 1.5 miles to the southwest. I continue past the left turn for Tom Dula Road and I cross the bridge over the Yadkin River and drive into the community of Ferguson. Soon I come to the bridge over Elk Creek and the sign that indicates that this is Elkville. There's not much there now. I turn the car right onto old Elk Creek Road, following the path Tom Dooley took on the night he left home headed toward the communities of Darby and Triplett and then on up to the crest of the Blue Ridge. Elk Creek Road is a lovely drive. It parallels the pretty Elk Creek through farmlands and wood, constantly climbing up to the old Boone Trail—about where the Blue

Ridge Parkway is today. The old road that runs beside the creek offers many gurgling cascades and pools to soak weary feet. But walking the road must have been long and hard—particularly hard and lonely for Tom Dooley, as he left his home and Ann Melton behind without knowing if he would ever see either again.

The old path Dooley took, Elk Creek Road, finally stops climbing where it runs underneath the Blue Ridge Parkway and then intersects Highway 421. Tom likely followed the old trails to Perkinsville and Sands on his way to Meat Camp. I turn left on 421 and drive to Perkinsville where I turn north on Highway 194. At Meat Camp Creek I turn left on Meat Camp Road and begin the climb up to Rich Mountain Gap. I turn left again onto State Road 1300 and continue climbing. In 1866 this was the main route to Zionville and Trade, Tennessee. The gravel road winds up between the great bulks of Rich Mountain (elevation 5,372 feet) and Snake Mountain (elevation 5,574 feet). I stop and park in the middle of the Gap. Much as Tom Dooley must have done 136 years ago, I gaze back down from where I've come—back at North Carolina. Then I walk over to the old road heading down on the other side of the Gap. I stare down into Tennessee—the same view that confronted Dooley as he contemplated what he must have thought was his successful escape. And the beginning of his new life.

Back in the car I head down the mountain into Tennessee. A flock of wild turkeys cross in front me. Further down the mountainside is all residential development. Then I am in Trade, Tennessee, just down the road from Zionville. Tom Dooley probably arrived in Trade on about July 2, 1866. In Trade, Dooley found work on the farm of Colonel James Grayson. Perhaps tipped off that he was being pursued, Tom left Trade after about a week and headed deeper into Tennessee. Deputies from Wilkes County met with Colonel Grayson on July 10 and together they set out in pursuit. The posse overtook Tom Dooley in Pandora, Tennessee, about nine miles west of Mountain City. The story goes that Colonel Grayson picked up a rock and told Tom he was under arrest. They took Dooley back to Trade, riding behind Grayson on Grayson's horse. Grayson and the deputies carried Tom back to the Wilkesboro jail on July 11, 1866. Tom would remain incarcerated from that time until his hanging in May of 1868. On 30

April 1868, the day before his hanging, with the knowledge that Ann Melton would be tried for the murder after his death, Tom Dooley wrote out a statement that he was "the only person that had any hand in the murder of Laura Foster."

From Trade, I turn back toward North Carolina on Highway 421. Soon I'm driving through Boone, and then turning under the Blue Ridge Parkway to head back down Elk Creek Road to Elkville. Arriving at the intersection of Elk Creek Road and State Road 268, I turn right on 268 headed away from Wilkesboro. Very shortly I cross the county line into Caldwell County. And soon, in a field on the south side of the road I see the lonely grave of Laura Foster, on the banks above the Yadkin River. There is a turnout on the north side of the road and a metal plaque dedicated to Laura's memory. After awhile I start up the car and continue on 268 through the Yadkin River Valley, also known as Happy Valley, until I reach Highway 321. Then I turn for home.

As the asphalt rushes by, my mind wanders back to Tom's tearful departure from Ann Melton and his long, solitary hike up Elk Creek Road. And I still envision him lingering awhile when he reaches Rich Mountain Gap, pondering what he had done down in Elkville, what he had left behind there, and what new life awaited him in Tennessee. I do feel some sympathy for the poor boy despite his unforgivable crime. Maybe, in part, it's because of the song's suggestion that Tom did indeed hang down his head and cry. Maybe he cried for Laura Foster or maybe for Ann Melton. Or maybe for himself. If it hadn't of been for Grayson, though, we'd never have heard the song.

For a detailed, factual account of the Tom Dooley story see John Foster West's *The Ballad of Tom Dula* (Moore Publishing Company, 1977, reprinted by Parkway Publishers, 2002).

The grave of Laura Foster

Retracing the Great Indian Trading Path: Occaneechi Town to the Trading Ford

On a warm, sunny morning in November, I back my SUV out of the driveway of my home in Durham and drive out of town on a northwesterly tack. At Hillsborough I cross over the Eno River and I pass near the reconstructed Indian village of Occaneechi. I make my way to the western edge of town on Dimmocks Mill Road. About a half mile outside of town I cross the bridge back over the Eno River. In mid-bridge, I glance to my right, up river, and spot the railroad bridge that spans the water just beyond the site of an old ford across the river. I know its there because I've followed the old road bed down to the river and seen where the old road continues on the other side of the ford. Over the bridge, where Dimmocks Mill Road bends slightly to the right, I shift into a higher gear, and drive onto the Great Indian Trading Path. It's been paved at this particular spot—or so I'm told. My plan for today is to follow the Path some ninety miles or so to the traditional lands of the Catawbas and to stop at the famous Trading Ford on the Yadkin River near Salisbury. Three hundred years ago, in February of 1701, John Lawson took about a week to cover this same route on foot. I hope to make it to the Ford in time for a late lunch.

    The Great Indian Trading Path was actually a series of paths connecting the area of Virginia in the vicinity of present-day Petersburg with the Catawba Indian lands near present-day Charlotte. The Path was in use by the native population of traders and travelers long before the appearance of European explorers and traders. Early maps of this area show the Path crossing the Yadkin and then cutting across present day Davidson, Randolph, Guilford, Alamance and Orange Counties. Early European explorers such as John Lederer in 1670, Lawson in 1701, William Byrd in 1728, and Bishop Spangenburg in 1752, wrote at some length about the Trading Path. Bishop Spangenburg contrasted traveling on the Path to traveling over the land beyond the Path: "[O]n the Trading Path . . . we could find at least one house a day where food could be bought; but from here we were to turn into the pathless forest." A teenaged Daniel Boone probably walked the Path with his parents in 1750 when the Boone family moved down from

Pennsylvania to make a new home in the Forks of the Yadkin. George Washington traveled on a part of the Path during his southern tour in 1791. The Path continued to be heavily used on into the 19th century, as horses, and later wagons, replaced foot travel as the common means of transportation. Some have claimed that the route of Interstate 85 from Petersburg to Charlotte, and in particular the crescent connecting the cities of Durham, Hillsborough, Greensboro, Winston-Salem, Salisbury and Charlotte, was dictated by the location of these population centers which grew up where they did because of the proximity of the Trading Path.

The construction of Interstate 85, various other roads, farms, towns and other developments, have removed much of what might have remained of the Path itself. I am driving on the small country roads to the south of and parallel to I-85, with names like West 10 and Bowman Road. These roads may well have been built directly on top, or just to one side, of the Trading Path. I look into the woods on either side of the road for signs of any old roadbed but see nothing to make me stop and investigate. But I do stop at the old cemetery at the church at Hawfields on State Road 119. Churches were often built along the Trading Path and an old cemetery may well be a clue as to the location of the Path. A portion of the path is said to be found in the woods adjacent to the Hawfields cemetery. Sure enough, in the woods adjoining the cemetery, at the edge of a plowed field, I find a broad, deep depression in the land. It runs straight down a slight incline, about twenty feet across and about five feet below the level of the surrounding land. The forest has grown up in the road bed itself, obscuring the road. I believe I am standing in a remnant of the Trading Path itself.

From Hawfields, I continue driving west/southwest along country roads that approximate the route of the Trading Path. In the small town of Alamance I pass an historical marker which says: "Trading Path: Colonial trading route, dating from 17th century from Petersburg, Virginia to Catawba and Waxhaw Indians in Carolina passed nearby." Although you would never find it on your own, the Trading Path does actually still pass nearby. Earlier in the fall I was fortunate to accompany Trading Path historian Tom Magnuson on a public hike he guided along a section of the Path that still exists between the Bellemont-Alamance Road and Great Alamance Creek. On this hike a group of twenty of us followed the well-defined road bed, though forested over with secondary growth hardwoods, for almost a mile to a ford of the creek. Magnuson, head of the Trading Path Association, told us about his study of the Path and efforts to preserve its heritage. Magnuson notes that locating the remnants of the Path often depends on discovering the river fords that provided the only passage across Piedmont rivers before the advent of bridges. "Find the fords and you'll find the footpaths. Find the footpaths and you'll find the villages," says Magnuson.

From Alamance I drive State Road 62 to Julian in the edge of Guilford County. This road may well follow the old course of the Trading Path. There is evidence of old road beds in the woods crisscrossing the right of way of 62. At several points I pull over, park and tramp through the woods to look at possible sites. They seem to be likely candidates but there is no way for me to tell for sure. Driving through Julian, however, I find myself on a secondary road that, according to the road sign, is named "Colonial Trading Path." This evidence may not satisfy an historian but it lets me feel that I am still on the right path.

I follow Old Red Cross Road and New Salem Road into Randolph County to Randleman—but past Randleman I am guessing. A Trading Path scholar from the 1950s, Douglas L. Rights, said that old maps show the Trading Path traveling though Randolph County's Caraway Mountains and that this little mountain range actually appears much as John Lawson described in his writing about the Path in this section of North Carolina. Rights also notes that the Keyauwee village where Lawson spent the night in 1701 could well have had its name transformed into the name Caraway over the centuries. I drive to Flint Hill and down Flint Hill Road through the Caraway Mountains. I find myself in agreement with Lawson and Rights—these are impressive mountains even if located in the Piedmont. I turn onto Highway 64 and immediately cross Caraway Creek with Shepherd Mountain and Ridges Mountain in the distance. Archaeologists have located the site of an Indian village in the bottom lands of Caraway Creek that they believe is Lawson's Keyauwee Town.

I cross the Uwharrie River on Highway 64 knowing that somewhere nearby is the Trading Path's ford across the Uwharrie—probably identifiable even today. I don't have the time to explore and, of course, private property rights must always be respected. I soon turn off Highway 64 following any roads that head due west. I may still be in the vicinity of the Trading Path but the roads I am following are now more questionable. High Rock Lake now covers the section of the Yadkin before me and I must turn northwest, away from the Path, in order to find a crossing of the Yadkin. I join Interstate 85 and drive 65 miles per hour for a short stretch before I exit on Highway 70 just before crossing the Yadkin. On the far side of the river is another historical marker. This one proclaims the "Trading Ford: On famous trading path used by Indians and early settlers. There Greene retreating from Cornwallis crossed on Feb. 2, 1781. East 1 mi."

In early February of 1701, John Lawson stayed several days at Sapona, the Indian town at the Trading Ford. Calling the Yadkin the Sapona River, he described the locale as follows:

> [W]e reach'd the fertile and pleasant Banks of Sapona River, whereon stands the Indian Town and Fort. Nor could all Europe afford a pleasanter Stream, were it inhabited by Christians, and cultivated by ingenious Hands. These Indians live in a clear Field, about a Mile square, which they would have sold me; because I talked sometimes of coming into those Parts to live. This most pleasant River may be something broader than the Thames at Kingston, keeping a continual pleasant warbling Noise, with its reverberating on the bright Marble Rocks. It is beautified with a numerous Train of Swans, and other sorts of Water-Fowl, not common, though extraordinary pleasing to the Eye. The forward Spring welcom'd us with her innumerable Train of small Choristers, which inhabit those fair Banks; the Hills redoubling, and adding Sweetness to their melodious Tunes by their shrill Echoes. One side of the River is hemm'd in with mountainy Ground, the other side proving as rich a Soil to the Eye of a knowing Person with us, as any this Western World can afford.

To reach the Ford, I drive into the edge of Salisbury and turn left or east onto Long Ferry Road. On Dukeville Road I turn left again and drive through a residential area toward the river. At the end of this road is the Yadkin and the Trading Ford. Douglas Rights tells us that although the backwater of High Rock Lake now covers the Ford, the islands in the Yadkin River that made up a part of the Ford are still visible in dry seasons. It has been dry this fall so I am optimistic. But at the end of the road is the large and imposing Dukeville power station with visitor parking and a chain link fence between the parking lot and the river. I park, find a gate in the fence and a short trail down to the Yadkin. I can see the islands across the still water in the mid-afternoon sunlight. I decide that I am standing close by the famous Trading Ford near where Lawson must have lingered three hundred years ago.

I wish I could see the Path's old roadbed angling down to a free flowing and reverberating Yadkin, but I'm content with what I have seen on my day's travel. I still marvel at the survival of the fragments of the Path's old roadbed that I did find and I hope there are many more remnants for Tom Magnuson to locate and preserve. And some day soon, maybe I can help search for the Trading Path's fords over the Uwharrie, Rocky, Alamance and Haw Rivers. But that day will have to wait at least a bit. I'm due back in Durham so I back out of the Dukeville power plant parking lot and head for the on-ramp to the latest version of the Great Indian Trading Path. Soon I'm traveling again at 65 miles per hour heading north on I-85. I'll make Occaneechi Town, present day Hillsborough, before dark.

A remnant of the Great Indian Trading Path can be seen along a section of the Poet's Walk, at Ayr Mount Historic Site, 376 St. Mary's Road, Hillsborough, North Carolina.

For further information about the Trading Path and related matters:
- The Trading Path Association website: *www.tradingpath.org*.
- John Lawson, *A New Voyage to Carolina* (UNC Press 1967) (originally published 1709)
- Douglas L. Rights, *The American Indian in North Carolina* (John F. Blair, Publisher, 1957)
- Douglas L. Rights, "The Trading Path to the Indians," *North Carolina Historical Review*, Volume VIII, page 403 (1931)

- Stephen Davis and Trawick Ward, *Time Before History: The Archaeology of North Carolina* (UNC Press, 1999).
- Stephen Davis, Patrick Livingood, Trawick Ward & Vincas Steponaitis, *Excavating Occaneechi Town: Archaeology of an Eighteenth Century Indian Village in North Carolina*, a CD-Rom (UNC Press, 1998)
- Occaneechi Band of the Saponi Nation Web Page (with information about reenactments at the reconstructed Occaneechi Village in Hillsborough): *www.occaneechi-saponi.org*
- N. C. Archaeology Home Page: *www.arch.dcr.state.nc.us*

Remnant of Trading Path

Live Oaks with First Names

The next time you head out to the beaches of Hilton Head, Kiawah or Bald Head Islands, take a moment, before you leave the mainland, to visit some ancient sentinels that have presided over coastal Carolina life for centuries. Live oak trees, so called because they remain green all year round, cluster along the east coast of the United States from North Carolina to Florida, and then along the gulf coast to Texas. Scattered along the Carolina coast are a few huge live oaks that somehow survived for hundreds of years while most of their kind and age succumbed to the ax, the cross-cut and chain saw. The oldest and largest live oaks are marked by huge branches which often spread out horizontally until they dip back down to the earth to return again to the soil. These trees can inspire.

The ferry to Bald Head Island leaves from the little town of Southport located about thirty minutes south of Wilmington by car, at the mouth of the Cape Fear River. The Southport area was inhabited by several Native American towns long before Smithville (Southport's original name) was established in 1792. And in Keziah Memorial Park, at the corner of Moore and Lord Streets, in downtown Southport, is a living product of the Native Americans' long vanished occupation. The Indian Trail Tree is an 800 year old live oak that long ago was pushed over on its side so that now the main thick trunk extends 8-10 feet horizontally before the trunk resumes its skyward growth. The young tree was bent over many centuries ago by the local Native Americans to serve as a guide to point travelers toward an important site — in this case possibly the local fishing grounds. Bending saplings to grow into and serve as trail markers was a common practice of Native Americans and similar Indian trail trees can still be found throughout the Southeast — there is said to be another good example on private property in North Raleigh and there was a well-known Indian Marker Tree in Gateway Park in southeastern Dallas, Texas. In Southport, the old Indian Trail is long gone, but the Indian Trail Tree still points the traveler toward the water.

In the 1830s and 1840s the lowcountry planters in South Carolina were growing increasingly incensed over federal taxes imposed on the export of their agricultural products. Their congressman, Robert Barnwell Rhett, regularly held forth on the unfairness of these Federal Tariffs. On 31 July 1844, as many as five hundred South Carolinians gathered under a two hundred year old oak in Bluffton, South Carolina — a center of commerce at the time — to hear Congressman Rhett speak on the inequities of the Federal Tariffs. And more significantly, Rhett made the case for South Carolina's secession from the Union. The local planters were convinced and the resulting "Bluffton Movement" focused the State's secessionists and led to South Carolina's becoming the first state to secede from the Union on 20 December 1860. The firing on Fort Sumter, on 12 April 1861, and the War Between the States would soon follow. The great live oak that sheltered the participants at the 1844 meeting, the Secession Oak, still stands. If you are driving down Highway 278 on your way to Hilton Head Island, turn right onto Highway 46 for the short drive into historic Bluffton. Stay on Highway 46 through Bluffton and just outside the town limits (on the Pritchardville side), you will see the Secession Oak — on the left hand side of Verdier Cove Road at Highway 46.

There is a long stretch on the two lane Bohicket Road that runs across John's Island leading to Kiawah Island where thick trunked live oaks line both sides of the road for several miles. The tree limbs spread out and meet above the asphalt, keeping the road in shadow even at midday. The size and placement of these trees indicate the long existence of this route — and the fact that these large oaks only remain along the road indicate the extent of the timbering over the years. But one special tree somehow survived away from the road — survived for what is estimated to be 1,400 years. Look for the sign for the Angel Oak, and turn onto the dirt road, Angel Oak Road, that leads to the entrance to the small park owned by the City of Charleston. Just behind the gift shop is a majestic live oak — 65 feet tall with a trunk at least 25 feet in circumference. The tree's huge branches, one almost 90 feet long, stretch out horizontally, some dipping back to the earth. The tree was already ancient when Abraham Waight received the property in a land grant of 1717. The plantation was developed over several generations of Waights until Martha Waight married Justis Angel in 1810. Despite several popular alternative explanations, this marriage was the origin of the title, the Angel Oak. This magnificent tree suggests the grandeur of the long vanished lowcountry forests of the Carolinas. Some speculate that the Angel Oak is the oldest living thing east of the Mississippi.

Once you have visited these grand trees, head on out to Hilton Head, Kiawah or Bald Head. But when you are out walking along the sand, watching the waves roll in and the sun settling down to the horizon, think about the Angel Oak and the wonderful stories it must surely tell, when the breeze rustles through its leaves, of human visitation to its shade and shelter over the last thousand years.

The Angel Oak

Sassafras Mountain

Once again I asked myself why. Why bother, why come out here so far, for what purpose? I had no valid answers. At first I had seen our venture as something akin to sport, something to be racked up as 'an achievement.' Then I had seen it as a way of indulging my pride, a frivolous vanity. Now I saw our enterprise as utterly meaningless or, at best, as an alibi to roam where few even had the chance to, with the added hope of discovering a little more about ourselves. —Michel Peissel, from his book ***The Last Barbarian***, discussing his expedition to locate the source of the Mekong River in Tibet.

After lunch at Big Mike's in downtown Brevard, we continued heading southwest on Highway 64 through Transylvania County. The clouds were close overhead. Solid and low. It never stopped raining. When we reached Highway 178 we turned left and drove due south into the little hamlet of Rosman. We crossed the bridge over the French Broad River and glimpsed an historical marker for the "Estatoe Path." The sign said it was the "Trading route between mountain settlements of the Cherokee and their town Estatoe, in what is now South Carolina" and that it passed nearby. I'd have liked to have looked for it—but there would have been be no enthusiasm for such a search in this weather. The road started climbing and we soon passed the Eastern Continental Divide. Then we entered South Carolina and headed downhill to the first thing marked on the map, Rocky Bottom. At the Bottom we turned left on the first road that looked like it went somewhere. The road sign said it was called "F. Van Clayton." It climbed and curved for several miles up past old paths and roads off in the woods. I leaned forward in the passenger seat, watching the windshield wipers rock back and forth and looking for roads off to the right. That is where I expected to find the short path to the highest point in South Carolina—Sassafras Mountain.

I guess it was in August of 1986 when the idea first occurred to me. I had just finished hiking the short but very steep trail up to the summit of Mount Elbert—at 14,433 feet above sea level the highest point in Colorado. Why not hike to the highest point in all fifty states of the United States, I thought. Most of the highest mountains or hills in each state ought to be easier than Colorado's highest point I reasoned. I'd already done Mount Mitchell in North Carolina (6,684 feet), Clingman's Dome in Tennessee (6,643 feet) and Mount Rogers in Virginia (5,729 feet) on various backpacking trips in the 1970s. That left just forty-six states to go!

When I got back from my Colorado vacation I mentioned my idea to several people. Someone eventually told me that they had seen a book on the summits of each of the fifty states. I dutifully located the book at an REI outlet and learned that hiking the fifty state summits was not a new idea by any means. People had been mapping, hiking and documenting their conquests of these summits for years. But the appeal of such a quest was not diminished—at least not for me. This was going to be a long term project, attempted only as it fit into other parts of my life's travels. If I met other high point baggers at the tops of some of these summits, all the better!

And it has been fun—slow, savored fun, but fun nonetheless. In April of 1992, my wife and two young sons made it to the top of the highest point in Florida. Britton Hill. 345 feet above sea level. It was easily several yards from the parking lot to the Geological Survey marker. I had to carry my youngest son the whole way. Both going up and coming down! If it hadn't been marked we never would have even known it was a hill, much less the highest point in Florida. Britton Hill turns out to be the lowest high point of all the states. Through the years I also climbed Wheeler Peak in New Mexico, Mount Mansfield and Mount Washington in Vermont and New Hampshire, respectively, and several other "peaks." And then there was a memorable August day in 1999, when a running buddy of mine helped me surmount the highest points of three states: West Virginia in the morning (Spruce Knob, 4,863 feet), and Maryland (Backbone Mountain, 3,360 feet) and Pennsylvania (Mount Davis, 3,213 feet) that afternoon. So I've now made the highest points in thirteen states. I may make Nebraska this summer if my wife has a conference in Denver. Panorama Point (formerly Mount Constable), 5,426 feet. The word is it is "up a slight hill" and has "a nice prairie view."

But back to Sassafras. My wife and boys aren't quite as excited about these high points as I am. They'll come with me but there is great danger that they will become surly and unmanageable if the hike leader hesitates too much. Or shows self-doubt. As the rain had not yet let up and as I wasn't entirely sure we were on the right road, I was aware of the shrinking window of opportunity to locate the Sassafras trailhead. We kept driving up and curving past old roadbeds angling off into the woods. But there was supposed to be a sign—it said so on the High Points web page on the internet that I'd looked up earlier in the week. Still, every time we'd turn a corner, we'd see the road curving ever upward. But then there was a parking lot and a sign. It said property of Duke Power Company and indicated an access point for the Foothills Trail. This was it. We parked and walked through the rain past a gate up the road. In less than 100 yards, there was the summit, straddling the border with North Carolina. The highest point looked to be at the location of a big pine tree—but this was probably in North Carolina and so didn't count. We located the South Carolina benchmark somewhat down the slope and took the obligatory photographs of the daring summiteers. Our view was only of the lowering clouds and the summit trees in the swirling mist. But our mood was upbeat. We had hunted for and captured our elusive quarry. And yes it was raining harder but we were just minutes from our vehicle. We turned our backs on Sassafras Mountain and skipped off down the hill through the rain.

Arriving back at the Grove Park Inn in Asheville that afternoon, a bellhop observed our hiking outfits and asked us where we had been that day. We said we'd hiked to the highest point in South Carolina. Taking us for some confused Floridian flatlanders, he said that we must mean North and not South Carolina—thinking we must have driven up the Blue Ridge Parkway to the famous Mount Mitchell just north of Asheville. No, no. Been there, done that, we said. The mountain was Sassafras and the Carolina was South, we told him. He'd never heard of it. Apparently, the allure of Sassafras was news to him.

Author's Note: *For a guide to the highest points in each state and reports by people who have recently summitted these points, visit an excellent web page, called America's Roof: Guide to the Highest Places in the U.S., at http://www.americasroof.com/*

Tom Fowler

You could also climb the 40 peaks in the Southern Appalachians that are on the current South Beyond 6000 list. If you climb these by an approved route, then you qualify for membership in South Beyond 6000, and will receive a patch and a certificate to recognize your accomplishment. For further details, see http://americasroof.com/6000.shtml

Summiting Panorama Point - the highest point in Nebraska

Monarchs of the Blue Ridge

Let's suppose you are a monarch butterfly flitting around the fields and forests of North Carolina on a hot summer day in late August. What are you likely to get the urge to do? Find some flowers and drink some nectar? Maybe. Hook up with some cute monarch flutterers of the opposite sex? Maybe. How about taking a road trip? A long road trip. Like maybe taking a couple of months to fly over a thousand miles to the mountains of central Mexico where you can spend the winter clutching a tree along with several million other monarchs who have also migrated down from the north? Bingo. In the late summer and early fall, monarchs across the United States start fluttering south on a migration that is simply amazing.

Although I've always liked butterflies, I really hadn't thought much about them since becoming an adult. So I was minding my own business on a hot day in July, reading a *Seasonal Guide to the Natural Year: A Month by Month Guide to Natural Events* in the Carolinas and Tennessee, by John Rucker. I had noted on page 70, that manatees once were common in the waters at the mouth of the Cape Fear River, and that a few American chestnuts still survive in North Carolina, at page 169. But I wasn't getting terribly excited until I reached page 228. And there it said that, at Milepost 415.6 of the Blue Ridge Parkway, "[t]housands of monarch butterflies pass through Tunnel Gap in September as they migrate to central Mexico." And it said no more. Well, okay. Little, fragile butterflies. Alighting on flowers and opening and shutting their paper thin wings. These guys are winging it from North Carolina to central Mexico? Like over a thousand miles or so? And they are starting off by flying through a 4,325 foot gap in the Appalachian Mountains? Hmmmm. Now that was something I needed to learn more about. And I really felt the need to spend a warm September afternoon sitting in the grass at Tunnel Gap, gazing east toward the North Carolina flatlands. And watching the monarchs flit by me on their long way west, and wishing them Godspeed on their incredible journey. Assuming it was all true, of course.

I had a couple of months until Tunnel Gap would be filled with migrating monarchs, so I had some time to research the matter.

I'd always known that monarchs are those bright orangey-yellow butterflies with black outlines on their wings. They are about 10 centimeters wide and weigh about half a gram. Little things. I knew they flitted about blooming flowers, and I guess I knew they ate nectar from the blooms. But I knew nothing about their travels. For a long time, science didn't know much more than I.

In the early part of this century, it was thought that the monarch butterflies disappeared from sight in North America in the winter by hiding in hollow trees or other sheltered spots—essentially hibernating. In general, butterflies were not thought to migrate. But some thought that the monarchs might travel to warmer spots to escape the winter's cold—but no one had ever found their overwintering territory. Then in the early winter of 1974, a volunteer researcher found several battered and beat-up monarchs on the side of a road west of Mexico City. Driving further up into the mountains, the researcher came upon a concentration of millions of monarchs clinging to all exposed parts of a grove of pine, fir and cypress trees. Further research revealed that about 14 million monarchs had gathered in these trees covering only a few acres at an elevation of about 10,000 feet in the Mexican state of Michoacan—about a three hour drive west of Mexico City. This was the long sought after winter home of the monarchs. The entire population of monarch butterflies that would repopulate most of North America in the spring and summer were roosting at this site (and a few other nearby sites). The monarchs would roost there all winter, attached to the trees, without eating, in a state of semi-dormancy. This huge concentration of roosting monarchs is said to be the "greatest butterfly show on earth."

Sometime in March, as the days grow warmer and longer, the monarchs stir from their winter perch, stretch their wings, and start their journey north. As they travel north, they mate and lay their eggs on milkweed plants—and only on milkweed plants, which are the only thing the newly emerged monarch caterpillars will eat. When these caterpillars turn into monarch butterflies, they will continue the monarchs' repopulation of their breeding range in North America—for the monarchs that overwintered in Mexico do not usually survive for long after beginning the spring migration. There will be several generations of monarchs in the spring and summer—the average life span of these seasons' monarchs is about six weeks. But these are the monarchs that will spread throughout the United States and southern Canada, traveling north and east as far as they can go until the milkweed supply ends.

The year's last generation of monarchs, i.e., those born in the late summer and early fall, are different from those born earlier in the year. These monarchs will have a life span of more than six months and they will not breed until the last month of their lives—in the early spring of the new year. Instead they will heed the shortening daylight and cooler fall weather, and they will turn their backs to the north to fly south (and west) to somehow find there way back to the few forested acres in the mountains of central Mexico—and maybe even to the same tree—where their great-great grandparents spent the previous winter. The complete monarch migration thus requires several generations. No individual monarch butterfly ever actually returns for a second winter in Mexico. But somehow the descendants always know when and where to go to complete their own leg of the migratory path.

How do the monarchs navigate? No one really knows. Do the monarchs follow the same routes during each year's migration? No one knows for sure but they do use the same overwintering sites year after year. How far can monarchs fly in a day? They probably average forty to sixty miles, although one tagged monarch butterfly was located 265 miles from where it had been released the previous day—monarchs may be able to use air currents at different altitudes to speed them on their way.

I looked for monarchs throughout July, August and the first half of September. I went to various gardens with blooming flowers, at all times of the day. I did not see very many butterflies at all. And I saw only one monarch—dead and its wings crumbling—blown up against the concrete curb at a rest stop on the Interstate. This was not good.

On the Internet I found several monarch web pages with reports of a brutal freeze that had struck the mountains of central Mexico in the previous January. Those who visited the overwintering sites soon after the freeze reported that the ground was littered with dead monarchs, several inches deep. Some estimated that 75% of the monarchs had died. The numbers of monarchs beginning the northward migration in March must have been significantly diminished. One naturalist commented that the best thing local folk can do to help the monarchs recover would be to plant milkweed for their eggs and caterpillars. I don't even know what milkweed looks like.

Armed with my new-found knowledge of the monarch, by the third week in September I was ready to make the long drive to Tunnel Gap. I picked a warm sunny day and left early. The radio played, Greensboro, Winston-Salem, then Statesville and Morganton rolled by. Up the long hill through Swannanoa Gap, I first saw clouds hanging over Asheville. I accessed the Blue Ridge Parkway near the French Broad River and headed up toward Mt. Pisgah, paralleling the Shut-In Ridge Trail. And the clouds began to close in. I turned on my headlights. Once on top of the ridge line, a steady drizzle began. I was driving through the clouds and at times the visibility was only several car lengths. I passed Milepost 410, then Wagon Road Gap. Then Milepost 415 and soon a sign for Tunnel Gap and the turnout on the left. The rain was misting and the clouds were lowering as I hopped out of the car in mid-September—just the time to part the waves of monarchs that would be riding the breezes up the eastern face of the Blue Ridge to crest Tunnel Gap and continue on their way to Texas and central Mexico. I stood in mid-Gap, facing down the slope, and let the rain mist on my glasses. Nothing moved in the sky. No orange and black wings opened and closed on the trunks of trees or beneath the sheltering leaves of the gap's vegetation. There were no butterflies.

I did recall from my research that monarchs don't like to fly in the rain—that they will hold onto trees or bushes until it stops raining. But this was mostly a misting rain, and there ought to be some monarchs in a hurry to get south and so willing to test the weather. So I kept searching the skies—and checking the trees and bushes for monarchs hunkered down, wary of the weather. I did see a sign that confirmed that monarchs did indeed pass through the gap on their way to Mexico. It showed the migration pattern and had a lovely photograph of a monarch. But it was the only monarch I saw at Tunnel Gap. The rain misted, the clouds lowered and I lingered at the gap. Even in the rain it was an engaging spot to look for butterflies. But there were none to be found that day. After a long while, I returned to the car, pointed it back towards the east, and started home. Half way down the parkway, the rain stopped and the clouds lifted. Just past Black Mountain the sun peeked out. It stayed sunny all the way back home.

Later that month, back at home and back on the Internet, I read a report of a monarch sighting on the coast of North Carolina from the previous March. This monarch could not have spent the winter in central Mexico and have made it back to North Carolina by March. There had to be some other explanation. The writers speculated that the monarch may have gotten lost or been blown far off course by bad weather, and had overwintered somewhere along the Carolina or Georgia coast. If the butterfly survived the North American winter, maybe clinging to some pine tree near Charleston, by March it might have been beginning its journey to the north—looking in vain for the monarchs from Mexico who would be ready to travel and mate with it. His or hers would have been a lonely trip though. Even if the Mexican monarchs had recovered from the disastrous freeze earlier in the winter, they would have been many hundred of miles behind, never to catch up to this monarch pioneer.

I hope it was just the weather that kept the monarchs out of Tunnel Gap on the day I went to meet them there. I hope it wasn't the big freeze or the timbering of their overwintering sites in Mexico that has reduced their numbers so that only a few flutter through the gap. I will return to the gap next September and bring a folding lawn chair and maybe a cooler with some margaritas mixed and chilled. Maybe I'll try to be there earlier in the day—and of course I will pray for sun and calm air. But first, this spring, I will get me some milkweed to plant in my garden—and, hopefully, I'll be able to personally observe at least a few monarchs as they head north to repopulate eastern North America before the weather turns cold. Godspeed.

Postscript: *In late February of 2003, my eye caught a familiar orange and black image in my morning newspaper; a photograph of a monarch butterfly resting on a milkweed plant. What's this, I wondered, an obituary notice? I read the headline with bated breath. "Monarch butterflies recover from freeze." Wow. The article said* that scientists were marveling that despite last year's deadly freeze, which killed hundreds of millions of monarchs, the fir trees of central Mexico were once more covered with monarchs. The survivors from last year's winter, started north last March and began the repopulation of North America—and apparently 2002's several generations of monarchs flourished. Several hundred million completed the fall trip back to Mexico—twice as many as the scientists had expected. A remarkable comeback for a remarkable little bug.

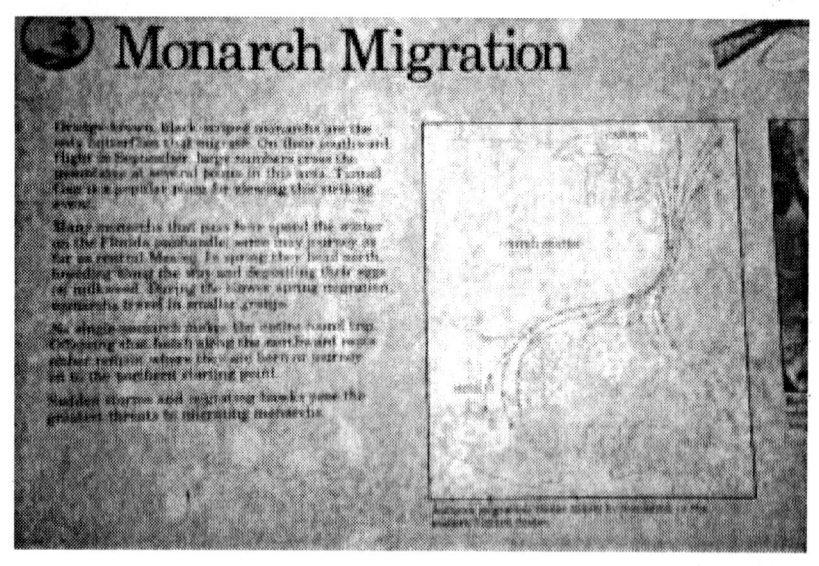

The Monarch Migration sign at Tunnel Gap

The Road To Adshusheer

Another Saturday morning and once again I'm standing in a jungle. In chest-high weeds. I can't even see my feet. It's a summer morning and I'm already soaked through with some combination of dew and sweat. The day started hot and humid and it's getting more so. Magnuson is nowhere to be seen. Or heard for that matter. I can't see any sign of an old road underneath all this sticky vegetation but twenty minutes ago Magnuson had said to push on across the little creek to the bluffs on the other side. Maybe we'd find the old road bed fording the creek and cutting into the bluffs, he'd said. Then he'd disappeared into the vegetation. And then again, maybe we wouldn't find the old road, I'd thought, as I moved off on my own. Maybe we'd just wander around in the weeds, swamp and muck, scaring all sorts of venomous snakes from their lairs, getting bitten by sullen, venomous brown recluse spiders, and working up a prodigious thirst. Maybe spending all these weekends looking for the path from Occaneechi to Adshusheer wasn't such a great idea after all. Maybe I was going to have a hard time even finding my way back to the car in these high weeds. And then, from far, far away, I hear Magnuson's shout. I can't make out what he is saying but his voice has that lilting note of ... well ... success? Victory? Has he found the old roadbed rising out of the bottom land and carving a v-shape ditch into the bluffs above the stream? I think he may have found it! I think we are on our way to Adshusheer.

In February of 1701, a young adventurer, named John Lawson, was somewhere in this same area—on his own way to Adshusheer. Lawson had traveled from Charleston, South Carolina, up the Santee River in a big dugout canoe, into central South Carolina. He then continued north to the area near present day Monroe, in Union County, North Carolina, where he had encountered the Great Indian Trading Path—the well-traveled trail leading north to Virginia from the lands of the Catawba Indians. Lawson followed the Trading Path across the North Carolina Piedmont until he reached the Eno River. He spent the night of February 12, 1701, in a Native American village called Occaneechi on the banks of the Eno near present-day Hillsborough. The next day, February 13, Lawson, with his new friend and guide Eno Will, left the Trading Path and hiked fourteen difficult miles to the east or southeast. That evening Lawson and Will spent the night in the Native American town of Adshusheer—a town that must be in or near present day Durham. Lawson's book, *A New Voyage to Carolina*, described the trip as follows:

> The next Morning, we set out [from Occaneechi], with Enoe-Will, towards Adshusheer, leaving the Virginia Path, and striking more to the Eastward, for Ronoack. Several Indians were in our Company belonging to Will's Nation, who are the Shoccories, mixt with the Enoe-Indians, and those of the Nation of Adshusheer. Enoe-Will is their chief Man, and rules as far as the Banks of Reatkin. It was a sad stony Way to Adshusheer. We went over a small River by Achonechy, and in this 14 Miles, through several other Streams, which empty themselves into the Branches of Cape-Fair. The stony Way made me quite lame; so that I was an Hour or two behind the rest; but honest Will would not leave me, but bid me welcome when we came to His House, feasting us with hot Bread, and Bears-Oil; which is wholsome Food for Travellers. There runs a pretty Rivulet by this Town.

From there Lawson and Will continued their trek east, passing by the Falls of the Neuse and probably by present-day Goldsboro, Grifton and Greenville. Lawson ended his 550 mile, 59 day loop through the Carolinas at the plantation of Richard Smith on the Pamlico River on 24 February 1701—where Lawson and Will parted ways forever. We know all this because Lawson wrote a popular book about his travels—the aforementioned *A New Voyage to Carolina*. Lawson later helped found two of North Carolina's oldest towns, Bath and New Bern, and, sadly, became the first victim of the Tuscarora War, when he was ritually executed by the Tuscarora near Contentnea Creek in September 1711. But that's another story.

As a long-time Durhamite, local history buff, and Eno River wanderer, I was well aware that February 2001 would mark the 300th anniversary of Lawson and Eno Will's trek through Orange and Durham County. I felt compelled to mark this historic anniversary—both personally and publicly. And it just seemed that the right way to mark it would be with a reenactment of the very trudge itself—the fourteen miles between Hillsborough and Durham. So in the spring of 2000, I put out the word for my fellow would-be reenactors and 300th anniversary celebrants: who would cross those "several other Streams" and walk that entire "sad stony Way" from Occaneechi to Adshusheer in February of 2001? Who would be the Eno Will to my John Lawson—or let me be Eno Will to their Lawson? Who would feast with me on "hot Bread, and Bears-Oil" once we shuffled in to Adshusheer? And where was Adshusheer anyway? Where was that "pretty Rivulet" that ran through the long missing town?

Left to my own resources, I probably would have simply followed the old fishermen's trails along the banks of the Eno River from Hillsborough to the Eno River State Park and called it close enough to Lawson's actual trail to count as a reenactment—but the fates had something else in mind for this 300th celebration. There was a late night telephone call. It was Magnuson. He'd heard of my reenactment idea and he had some ideas about the path Lawson may have followed and where Adshusheer could be found. Did I want to go take a look sometime? But of course.

It is quite amazing how many old roadbeds crisscross the woods and fields around Hillsborough and the land northwest of Durham. Interstate 85, Highway 70, Old 86 and Cornwallis Road all have changed the landscape and disguised the old roads and trails—but sometimes running parallel to these new roads and sometimes running off into the woods where the new road curves, those old road beds can still be found. And there is a guy in Hillsborough, Magnuson's his name, who knows where these old road beds are ... or where they were ... or where they should be. Like in snake and spider infested jungles covered with chest high weeds.

Tom Magnuson, local adventurer and backwoodsman, is the founder and CEO of the Trading Path Preservation Society. His organization is dedicated to "preserve, promote and study the historic Trading Path." As a result Magnuson is constantly researching the old road beds, trails, fords and tavern sites that cover the terrain around the oldest towns in North Carolina's Piedmont. A stickler for observing people's privacy and property rights, Magnuson always knocks on doors and tries to track down the owners for permission to follow these old roads and paths. The man knows his public relations and he knows most folk are deeply interested to know that hundreds of years of history happened on or near their property. Most folk are happy to have Magnuson investigate these old roads. And many have stories to tell. Magnuson hears about arrowheads and grinding stones found, old fords and fishing weirs, old gravesites and old stone chimneys standing by themselves in the middle of a mature forest. Then Magnuson tries to fit it all together.

    Magnuson tells me about his explorations and that he has several ideas about the old path that Lawson may have followed after leaving Occaneechi. He tells me the UNC archaeologists have never been able to definitively locate the site of Adshusheer. He thinks it's not to the north of Durham but west—and maybe even southwest of Durham. He knows the names of all the streams that might qualify as the pretty rivulet that runs by Adshusheer. Adshusheer might be near the intersection of Cornwallis Road and Old Erwin Road—or it might be on the edge of the Duke golf course. I'm entranced and, at least in my mind, Magnuson and I become a team for this reenactment hike business. We'll find a plausible trail, secure permission from all property owners for the reenactment hike, lead scores of Lawson/Will enthusiasts on the 300th anniversary hike in the clear winter air of February 2001, and finally we'll uncover the true location of Adshusheer and be awarded honorary graduate degrees in archaeology from that little college down in Chapel Hill. This would be fun.

    So, in the late summer of 2000, Magnuson and I do the legwork part of our little scheme. We meet on most weekends to explore undeveloped areas southeast of Hillsborough where he thinks there should be old roadbeds running southeast/northwest. Sometimes we wander the woods, swamps and streambeds for hours, and find nothing but stickers, bugs, barbed wire, and, afterwards, a numerous assortment of bug bites, scratches and rashes. Once my ankle, with several prominent red spots on it, swells up for two weeks. But we keep going back to the steamy Orange county jungle land, because sometimes we find the old road bed and the wide u-shape ditch that runs down to the stream, fords the stream and runs up the other side. And sometimes that sunken roadbed keeps on through the woods until it is a twenty foot wide roadbed set five or six feet below the level of the surrounding forest. To reach that depth the traffic on these old roads must have been heavy indeed. Then we follow the roadbed to where it disappears because it has been leveled by a modern road or house, barn or parking lot. Then, like hunting dogs, we spread out on the far sides of the modern developments to see if we can pick up the old trail, that is find a continuing remnant of the old roadbed. Sometimes we do and sometimes we don't.

But by early September it all seems to have come together—we have established an intermittent but almost continuous trail of old roadbeds and fords leading from the Eno River at Hillsborough to Mt. Sinai Road in Durham County. Almost there. The last section Magnuson and I find on one of the first fall-like weekends in September. Where Mt. Sinai Road curves off to the left down a hill, we turn into the woods on the right—with Magnuson guessing that the old roadbed on which Mt. Sinai was built may have gone off to the right instead of to the left. We wander along the edge of the hill, trying to stay level as the old cart way also might have tried. And then we spot a clearly defined ridge of earth running in a line heading southeast. It has to be the edge of an old roadbed. We follow it for 100 yards until it begins to head down the slope of the hill, and immediately it deepens into an eroded ditch about four feet below the level of the surrounding land. This erosion often happens when old roadbeds descend the bluffs surrounding streams. Over bushes and fallen logs we climb down the ditch until the old road flattens out on the small floodplain of Piney Mountain Creek. And there before us is a beautiful little ford of the creek—with the old road clearly rising up the bank on the far side of the stream. We wade the stream and follow the deep cleft of the old road as it runs through the mature forest gradually ascending. We pass a few old home sites and stone foundations and then the old road ditch runs perpendicular into Erwin Road—a hundred yards or so west of Hollow Rock Swim Club. On the far side—the south side—of Erwin Road, the cleft of the old road keeps on going until it merges with the gravel Pickett Road. Adshusheer may just be down this dirt road a ways. It could be, Magnuson, it could be.

But confirming Adshusheer's true location would have to wait. We had a hike to organize, authorize, publicize, and energize—and only until February to do so. So we got busy. Magnuson got permission from the property owners for our passage along the proposed trail, we organized committees to take charge of the starting point and finishing points of the hike, we got assistant hike leaders and support personnel lined up. And in late January of 2001 we completed the first continuous hike of our educated guess as to Lawson's actual sad and stony fourteen mile path from Occaneechi to Adshusheer. We forded the Eno just downstream from the Churton Street Bridge and made our way up to Tuscarora Drive, walked briefly on Highway 70, Highway 86 and Old N.C. 10, before turning off into the woods on the old roadbed. We lost our way twice (despite having been over the various segments of the trail several times before) but completed the path in about seven hours. When we emerged from the woods triumphantly on Erwin Road, we unanimously declared that the hike had been just under fourteen miles—supporting our belief that we had found the actual path Lawson and Will took in 1701. That was good enough for us and good enough, we felt, for the reenactors that we expected would join us for the 300th anniversary hike on 17 February 2001.

It poured rain the night of 16 February 2001 but on the morning of 17 February the clouds were scudding and by 8:00 a.m. the sky was clear and bright. After registration and opening remarks in the parking lot near the reconstructed Occaneechi Village on the banks of the Eno in Hillsborough, about 75 hikers started out on Lawson's "sad and stony way" to West Durham and at least the possibility of Adshusheer. The day was glorious and long, and the hikers spread out along the trail. But all hikers had reached the finish at the Hollow Rock Swim Club parking lot (or were otherwise accounted for) while there was still plenty of light left in the day. And all felt a bit closer to those fellow wanderers who slept well in the village of Adshusheer on the night of 13 February 1701.

Postscript: After the reenactment hike, Magnuson made the following observations about the hike: "First, and most amazing, was the miracle of over 60 property owners granting permission for strangers to trek across their land. Second, and nearly invisible to the hikers, the success of the venture depended entirely on a cadre of selfless volunteers. For the record, the volunteers who made this hike possible were: Diane Magnuson, Holly Reid and Rich Shaw, Bill and Gwen Reid, Tom and Gail Fowler, Steve Rankin, David Southern, Bryan Carey,

Gordon Warren, Cindy Shaw, Annette Jurgelski, Walter Rogan, Eric Block, Robin Jacobs, Milo Pyne, Peter Klopfer, Gustavo Ocoro, the Occaneechi Band of the Saponi Nation, and Chris Pope and Kent McKenzie of Orange EMS. Furthermore, the hikers in particular and the community in general owe a special debt to the wonderfully generous but necessarily anonymous landowners who share their piece of the rock with us. Several segments of the hike are in the hands of extremely responsible landowners fully aware of the wondrous treasure they possess. The owners, who weren't aware before the hike, were afterward, and perhaps that new knowledge will provide a degree of protection for a wonderful artifact in our midst. Thank you all for contributing to a moment of grace none of us will soon forget." And, of course, thank you Tom Magnuson for all of your crucial work on this hike and your tremendous efforts with the Trading Path Association!

And as for Adshusheer? I have a clipping of an article which appeared in the *Durham Morning Herald* on Sunday, 3 December 1989. The headline reads: "Durham Site May Yield Sought-After Indian Village." The article discusses an archaeological site discovered by an amateur archaeologist that had attracted the interest of the university archaeologists. The article described the location of this site only in general terms: "on private property beside a creek within the city limits—to avoid drawing unwanted attention to the site." The article continued: "What is particularly tantalizing about any site in Durham County is the fact that it could prove to be Adshusheer, a village described by English adventurer John Lawson in his book A New Voyage to Carolina, published in 1709. The settlement, inhabited by Eno and Shakori Indians, disappeared from history and has never been found by researchers. 'We've been looking for that village for a long time,' [said one of the university archaeologists]." But despite the great interest of both amateurs and professionals and the discovery of possible sites, no one has announced publicly that they have found Adshusheer. Magnuson stills says Mud Creek (near Cornwallis and Erwin Roads) might be the pretty rivulet. But Eno Will's Adshusheer still remains buried somewhere in Durham, enduring the centuries and eluding our grasp.

Visit Magnuson's Trading Path Association web page on the 'net at *www.tradingpath.org*

Tom Magnuson searching for old roads

Old road leading toward Adshusheer

The Shut-In Trail

George Washington Vanderbilt was a rich man. In 1888 he began to buy up tracts of forest south of Asheville. He bought up the land surrounding the long ridge that runs up from the French Broad River to Mount Pisgah. And then he bought the top of Mount Pisgah itself from Thomas Lanier Clingman (of Clingman's Dome fame) who had owned the property since the 1830s. In addition to the famous 255 room Biltmore House, Vanderbilt constructed a log hunting lodge at Buck Springs Gap near the top of Mount Pisgah. He called it Buck Springs Lodge. To get from Biltmore House to Buck Springs Lodge, Vanderbilt needed a trail. So he built one that ran from the Biltmore House up the ridge from the French Broad to Buck Springs Gap. Probably because of the thick rhododendron thickets covering parts of the ridge, he called it the Shut-In Trail. And it's still there.

The Blue Ridge Parkway and the Shut-In Trail now weave back and forth across each other, roughly parallel, as they climb up to Mount Pisgah. But the Shut-In Trail was there first — the Parkway wasn't built until the 1930s. The Trail now starts at the N.C. 191 off-ramp of the Blue Ridge Parkway at the Bent Creek Parking Area on the French Broad River. The trail is about seventeen miles long, starting at an elevation of 2,025 feet and ending at the Mount Pisgah parking area at about 5,000 feet (near milepost 408 on the Parkway). But because the Trail contains lots of downhill sections, the Shut-In traveler ends up climbing much more than 2,975 feet.

There is a running race on this Trail, held in November every year since 1980. A race brochure from this Shut-In Ridge Trail Run describes the course, as follows:

> [I]f you have run this race before, you might recall what it is like, know what you are getting yourself into, and can disregard the following if you wish. If, however, time has eroded your memories, or if you are a 'First-timer', I hope you are truly aware of the nature of this course. It is almost 17 miles which is run, walked, and even crawled. The uphill portions total more than 5,000 feet with a net gain start-to-finish of almost 3,000 feet. There are long, steep climbs — some with log steps and switchbacks. There are also several equally steep downhill sections (about 2,000 feet worth) where a fall could easily result in an injury.

The footing is rough, leaf-covered at this time of year — with roots, stumps, uneven terrain, and several very rocky sections. BE CAREFUL!! Since there have only been a dozen or so people who have been able to run this race at a pace quicker than 9 minutes per mile, it might be a mistake to 'attack' this course from the gun, even with the 'easy start.' Saving a little for the last 2 miles from Highway 151 to the finish is advisable.

For many of us, who were running at the time of Frank Shorter's Olympic marathon victory in 1972, running a marathon became the goal for the remainder of the 1970s. But in the 80's, we'd done the marathon thing and were ready for the next thing. And it was cross-training, it was triathlons, and then, it was trail running. And trail running, in North Carolina, meant signing up for the Shut-In Ridge Trail Run. As always, you send in your application and entry fee months in advance, then hope your training goes well and that you stay injury free for those crucial months leading up to the race. In 1987, after years of talking about it, five from our local running club sent in our Shut-In applications. Soon thereafter, we five received confirmation that we were officially in the race.

So we trained, as best we knew how, and on the first Saturday in November we showed up, along with 150 or so other aspiring trail runners, at the Bent Creek Parking Area. The runners milled around, eyeing each other and talking about what clothes to wear or not wear. The French Broad flowed majestically by. We crossed N.C. 191 to the old dirt road and lined up at the starting line. At 11:00 a.m. someone said "Let's go," and the Shut-In Trail Run had begun.

We all knew that few had completed this race in less than three hours and that a ten minute per mile pace would bring you in well under three hours. In 1987, a ten minute mile pace seemed very manageable — particularly at the beginning of the Shut-In Run as we trotted along the relatively flat dirt road. After less than a mile, the course turned off the road onto an actual trail through thick rhododendron. The trail began a significant climb with several switchbacks. The runners began to slow down and back up. We couldn't pass in places and were forced to walk behind the slower runners. But the narrow trail soon turned onto an old forest road and although we kept climbing, we could settle into a comfortable pace.

The day was sunny and relatively warm. This wasn't going to be so bad.

After climbing for awhile we hit a number of flat stretches. There were mile markers and our watches showed we were staying well under nine minute per mile pace. Our five runners were staying fairly close together, often seeing each other at the points where the trail intersected parking areas on the Blue Ridge Parkway where the race organizers had their water stops. First Walnut Cove Overlook and then Sleepy Gap Overlook. Inevitably, after these parking lots and water stops the trail would begin again with a steep climb over some knob or ridge. We began to see a lot more walkers than runners — at least up these steep sections. There were long sections with only gentle climbs as we traversed heavily wooded slopes. The roots and rocks on the trail were not too bad. Our group started to get spread out — unfortunately with me being one of the ones to start falling back.

Just after eight and a half miles we hit another parking lot and water stop — Bent Creek Gap. I wasn't feeling awful but I was the last of our group — and I wasn't passing anybody anymore. Maybe this wasn't going to be a romp to the summit of Mount Pisgah, after all. Immediately after this water stop, the trail picked up an old mountain road that climbed ungently across a ridge and then switchbacked further up the mountain. This road was rocky, long and depressing. Almost at the top the road turned straight up the mountain and I strode weakly up toward the summit with my hands pushing down on my knees at every step. Then I was at the grassy summit of Ferrin Knob at 4,064 feet — a pretty spot but I was none too encouraged. It was about the ten mile mark — seven miles to go and I was dragging. And worse, the trail immediately went into a steep and lengthy downhill section. The footing was awful, yet runners bounded by me in a gravity-induced euphoria. And we were loosing all that altitude that had been so preciously gained — altitude that we would have to preciously regain, I knew. I was behind all my running buddies and continuing to loose ground, as far as I knew. Maybe this wasn't so much fun after all. But there was nothing for it but to keep going.

And sometimes, when you are plodding along in life or on a forest trail, with no particular expectations — certainly no high expectations — except that time and distance will pass, you find, to your surprise, that several miles have gone by and you're feeling pretty good. You look up ahead and there is one of your running buddies shuffling along. You catch him and say, "Keep it up, Doug!" and slowly pull away. The uphills don't seem so steep anymore and you start passing other runners. And another one of your friends. You are upbeat and encouraging, and you are clearly picking up the pace. You possess no explanation for this turn of events — but you'll take it. After the Stoney Bald Overlook on the parkway, there is a long relatively level stretch of trail. It's rocky, rooty and leaf-covered. But you still feel good and continue to catch and pass other runners. As you come into sight of the last water stop, Elk Pasture Gap (4,200 feet) at about the fifteen mile point, you have caught all but one of your running buddies, and you know you will finish. You stop and take thirty seconds to down a couple of cups of water and whatever else they are offering. Then you start the last leg.

The last leg of Shut-In is, however, a bear. Although it is only 1.8 miles and there is a flat, grassy section, most of it is straight up the mountain — no carved switchbacks, no following contours, just the steepest most direct way up. The trail toys with runners' thighs, breath and determination. Strong runners, who have averaged faster than ten minutes a mile so far in the race are staggering or sitting on rocks, waiting for some sign ... or some reason to continue. I'm struggling upward, not stopping but moving in slow motion, and watching the sun shining down at me from just above the top of the ridge line. As I squint and grunt, the sun is momentarily blocked by something further up the trail. Continuing up, the sun reappears behind a runner holding on to a tree at the side of the trail. It is Rob, the last of the running friends with whom I had signed up for the Shut-In Run so long ago. Rob, a mentally tough, veteran racer, is hanging onto the tree as if afraid he is about to slide down the slope to oblivion. I'd like to stay and talk but I am afraid to stop. I nod at Rob — and I think I smile — and move, agonizingly slow but still steady, up the mountain. We're all on our own the rest of the way at Shut-In.

Over this last section, I'm managing a 16-17 mile per minute pace. Not that I care at this point. Finishing is the only goal. And at some point I realize that I'm no longer climbing. And then I hear voices. The last hundred yards of Shut-In are strange. The trail runs downhill on a very rocky and rooty route that allows no view of the finish. You can hear voices but only at the very end, when the trail takes a sharp right turn down some stairs, do you see the people, the finish chute and the parking lot where you can take a load off and pull off those shoes. It is a wonderful sight. And if you position yourself just right, as your legs stiffen and you sip your well-deserved beverage, you can see your running buddies emerge from the woods, pick their way down the few steps and finish their first Shut-In Ridge Trail Run. And you can muster the energy to yell appropriate commendations as they make their way to collapse on the grass next to you.

I've run the Shut-In Ridge Trail Run several times since that first one in 1987. Sometimes the weather has been quite cold and sometimes quite warm. Once a snow forced the run to end after only eleven and a half miles. But I've never run it as fast as I did that first time — maybe it's because after the first time I've always known what to expect from the trail and so have erred on the side of caution in setting a pace. But of course there are other explanations. The last time I ran Shut-In was also my slowest. My wife and two young sons served as my road crew — driving up the Blue Ridge Parkway, stopping at the water stops to meet me with drinks and encouragement. As they worked their way up the parkway and saw all the runners go by while they were waiting for me to appear out of the woods, my boys started getting a little bored. And immediately after I would drink their drinks and head back to the trail, they would drive off to the next water stop — never seeing if there were any runners behind me. So at the finish, they watched the many runners emerging from the trail and they waited to see me. When I finally appeared and crossed the finish line, my boys cheered and then my eldest son (seven years old at the time) turned to his mother and asked: "Mama, was Papa last?" Well, not quite, but I saw his point.

The finish of the Shut-In Ridge Trail Run

History on a Stick

I'm cruising down a lonely stretch of Highway 731 in Montgomery County. I've passed the turn off to Town Creek Indian Mound. There seems to be no one else driving this road today and not many houses built along this way, either. I'm driving the speed limit having taken the long way home—just for a change. I don't think I've ever been on this stretch of road before. And then suddenly, a familiar black on silver shape beside the road catches my eye. Way out here in the middle of nowhere, it's a state historical marker perched on a metal post five feet off the ground right where the road crosses a small creek. I know at my speed I'll never be able to read more than the first couple of words, so I glance in the rearview mirror (no one there), downshift and brake. I pass the marker, do a three point turn and pull off the road in front of the sign. I am much surprised by what I read. It says: "Flora MacDonald: Scottish heroine who lived in N.C., 1774-79. Loyalist in the Revolution. Her home stood on this creek a few miles north."

Well, that hardly does the Flora MacDonald story justice, I think. I had recently read up on Flora MacDonald's famous escape with Bonnie Prince Charlie from the bloody Duke of Cumberland after the Scottish defeat at Culloden in 1746, and Flora's subsequent unhappy immigration to and sojourn in North Carolina during the tumultuous Revolutionary War period. (A truly fascinating story, by the way). I had also recently visited the Cross Creek home site (in present-day Fayetteville) where Flora Mac had lived when she first arrived in North Carolina. So I knew at least a part of the rich story that dwelt between the lines of the terse, twenty-one word inscription on this historical marker. And I would have happily walked a mile or two along the little creek that crossed Highway 731 if I could be sure to stand on the site of another of Flora MacDonald's homeplaces. But "on this creek a few miles north" wasn't going to help me much to find the precise spot where Flora Mac lived. And I don't think it was ever intended to The placement of the marker on this secondary road, miles from the actual house site, with a twenty-one word summary of Flora's significance and no more direction—well, the marker was clearly just a tickler, just a tease. As usual, the historical marker gave just enough information to engage the interest of someone already interested but no guidance as to what to do next to pursue this interest. As usual, the historical marker left me feeling somewhat conflicted and unfulfilled. Maybe a bit like Shelley's traveler from an antique land who, after stumbling upon Ozymandias' ancient historical marker, looked around in vain for the grand "works" referred to on the stone pedestal, but instead found that: "Nothing beside remains. Round the decay of that colossal wreck, boundless and bare, the lone and level sands stretch far away."

Like most of the other states, North Carolina's historical marker program is long established. As detailed in the recently published Ninth Edition of the *Guide to North Carolina Highway Historical Markers* (published in 2001 by the State Division of Archives and History, edited by Michael Hill) North Carolina's first historical marker went up on January 10, 1936, in Granville County, telling us, in seventeen words, about one John Penn, a signer of the Declaration of Independence, who lived "3 miles northeast" of the marker. Sixty-seven years later, there are now 1,434 historical markers scattered across our state, each supposedly marking a spot "where history happened," as the *Guide* tells us. Each of our one hundred counties has at least one historical marker with Wake County having the most—seventy-two. But in reality the history that really happened, at least for many of the historical markers, happened at a spot physically distant from the historical marker—sometimes a matter of miles. For instance the Great Indian Trading Path has ten historical markers spread across the state but none are actually on the Path and all simply conclude that the Path "passed nearby."

According to the *Guide*, a person or event gets a historical marker only after being adjudged of "statewide historical significance" by the Highway Historical Marker Advisory Committee—members of this ten-person committee are usually history professors appointed by the secretary of the Department of Cultural Resources. The Committee also approves the twenty or so words that summarize the significance of the person or event that appears on each marker. The Committee's guidelines state that the inscriptions on the markers shall include only straightforward, undisputed historical facts—not that one could do much editorializing in the twenty or twenty-five words that will fit on the historical marker's few available lines of text. In any event, the guidelines also state that "[w]ords such as 'great,' 'outstanding,' 'important,' will not be included in marker inscriptions." The markers are never placed on the Interstate Highways but are generally found on the right of way of the state's secondary roads. The *Guide* tells us that to increase their chances of being read by motorists as they speed by, markers are often located "at a turn-off or at an intersection where traffic slows." Maybe so, but many is the time I've been unable to read the entire marker's text before the sign flashes past—and this at speeds well below the speed limit.

North Carolina's legislators are not indifferent to the factual accuracy of these historical markers. A special statute provides for a procedure for a citizen to challenge either the erroneous placement of a marker or erroneous information contained in the marker's inscription. But then again, a number of markers purport to track Hernando de Soto's trek through North Carolina in 1540, despite no general agreement among the experts as to de Soto's actual path. And anyway, how can anyone prove that the de Soto historical markers of Jackson, Macon, Cherokee and Clay counties are erroneously located when they only proclaim that de Soto's expedition "passed near here." Of course, I only wish I knew enough of Flora Mac's interlude in North Carolina to know whether or not she indeed lived for a spell "a few miles north" of the historical marker on that lonely stretch of Highway 731. No, the Highway Historical Marker Advisory Committee probably doesn't lose many of these challenges to its markers' placement or accuracy.

And just because it's interesting, you should know that the relevant legislation does restrict use of these historical markers to only those expressly approved by the Committee ... but if you surf the Net—looking for key words like "historical marker"—you can find the following: "Get your own State Historical Marker to place in your yard or home. Looks just like the real thing but in place of a Historical Event, a hilarious, make no sense saying is displayed. Make your house a hilarious historical site from your state. $40.00 each." The site gives an example for the state of Utah—the marker reads: "On May 10, 1869, East met West as the last spike was driven, completing the Union Pacific Railroad. Many years later, the last nail was driven, completing this building." Boom, bada, boom!

But, okay, despite its limitations, the historical marker program does serve a noble purpose, at least in theory. Undoubtedly many North Carolina travelers are not going to take the time to read the *WPA Guide to the Old North State*, Daniel Barefoot's touring guide of North Carolina's Revolutionary War sites and Clint Johnson's guide of the Civil War sites, or William Powell's histories of North Carolina. For those busy travelers, their knowledge and appreciation of the people, places and events that have shaped North Carolina may be limited to the parts of these historical markers that they are able to read (assuming they make the effort in the first place) as they speed on their hurried traveler's way. And that, by itself, may be worth the program. When they see in Craven County the marker for "Bayard v. Singleton," a famous North Carolina Supreme Court case, or in Henderson County the marker for "Wolfe's Angel," the marble statue that inspired the novel *Look Homeward, Angel,* or in Green County the marker for "Nooherooka," a Tuscaroran Fort where the decisive battle of the Tuscaroran War was fought, or in Durham County the marker for the "Bull City Blues," a tribute to bluesmen Blind Boy Fuller and Gary Davis, maybe some of those in their automobiles are momentarily transfixed by thoughts of these significant but long disappeared people and events. And even if it is only momentary, that brief transcendence of their actual, present trip to the beach, the mountains, Carowinds, or even a state historic site ... well, that transcendence must be a good thing.

And, to be honest, the same applies to me as well. I've seen all the "Stoneman's Raid," "Cornwallis," "Juan Pardo," and "Regulator" markers, so often that I rarely bother to read the text as I drive by anymore—but then, when I least expect it, on a lonely road in an unexplored county I stumble on a marker for Flora MacDonald, and, involuntarily, I pause, both physically and mentally, and am transported, at least part of the way, back in time. And I imagine Flora Mac languishing in the oppressive summer heat of the Carolina Piedmont in a modest home on the banks of a small sluggish creek, and dreaming of the cool, rocky coast of Scotland's Isle of Skye and the bonnie prince, awkwardly disguised as her servant girl, "Betty Burke," as they approach the British patrols bent on their discovery and capture.

Wolfe's Angels

I lived in one of the fourth floor attic rooms of Mangum Dorm my freshman year at the University of North Carolina in Chapel Hill. It was a memorable year, of course, as all freshman years are wont to be. I struggled with calculus, was only mildly interested in chemistry, and was annoyed at the huge introductory courses in political science, anthropology and psychology. But I was amazed at the freedom I had. Freedom to stay up late, walk around campus or downtown to Franklin Street at any time of day or night, and, of course, much more. Attending eight o'clock classes came to seem a physical impossibility. Sure, the alarm would go off but I would immediately fall back asleep and wake again only when the radio shut off after the thirty minute alarm mode expired. But my sleep deprivation wasn't all because of late night socializing and beer tasting. Many were the nights I was awake at 1:00 in the morning, sitting at my desk in my Mangum attic room, reading. But it wasn't homework.

Certain novels I stumbled upon simply swept me away—I hate to use the cliché, but I literally could not put these books down, and read them into the nights and early mornings. Kerouac's *On the Road*. John Barth's *The Sotweed Factor*. Robert Penn Warren's *All the King's Men*, and Kesey's *Sometimes a Great Notion*. Calculus and chemistry be damned, my need to remain in the worlds created by these novels was too great to be displaced by coursework or sleepiness. And maybe at the top of the list of these books that spirited me away my freshman year was Thomas Wolfe's famous autobiographical novel of growing up in Asheville, *Look Homeward, Angel*.

Before it burned, I visited the boarding house in Asheville run by Wolfe's mother and described at length in *Look Homeward, Angel*, called "My Old Kentucky Home." At the time the keepers of the house had posted descriptions from the novel of various parts of the house at the described location within the house. The effect was remarkable—to read Wolfe's words describing the staircase and then to look up at the actual staircase itself. Wolfe and his fictional counterpart from his novel, Eugene Gant, were also freshmen at the University of North Carolina—in Chapel Hill for Wolfe, and in "Pulpit

Hill" for Mr. Gant. As an undergraduate, many is the time I strode through Battle Park or Polk Place in their footsteps. It wasn't until recently, however, that I learned of an important *Look Homeward, Angel* site that I'd never visited—the homeward looking angel itself.

Thomas Wolfe's father was W.O. Wolfe (W.O. Gant in the novel) who made his living as a stonecutter and seller of tombstones. On the porch of the elder Wolfe's shop in Asheville, Wolfe kept one or more carved marble angels that he had ordered from Carrara, Italy. The stone angels apparently sat on the porch for years, unsold and staring off into the distance—making an impression on the youngest of W.O. Wolfe's eight children—Thomas. *Look Homeward, Angel* contains several references and descriptions of a stone angel, resting for years on W.O. Gant's porch, "poised clumsily upon the ball of one phthisic foot, and its ... white face ... [wearing] the look of some soft stone idiocy." Wolfe's angel is still with us.

Twenty-five miles south of Asheville (Eugene Gant's Altamont) in Hendersonville is the Oakdale Cemetery—from downtown Hendersonville take Highway 64 west a short distance. You'll see the cemetery and then on your left you'll see a state historical marker which states: "Wolfe's Angel: Marble statue from the Asheville shop of W.O. Wolfe inspired the title of son Thomas Wolfe's 'Look Homeward, Angel.' Stands 150 feet south." Sure enough, just to the south is a wrought iron fence surrounding the grave of Margaret Bates Johnson. On a pedestal above the grave is a stone angel with raised arm and a beatific expression. This is likely the angel that Thomas Wolfe knew well as a child, surely studying its expression and following its gaze as he passed by it on the porch of his father's shop.

But Hendersonville's claim to Wolfe's angel is not without a little controversy. It does seem probable that W.O. Wolfe imported several marble angels from Italy and that more than one may have lingered at his shop during Thomas' youth. There are reports that W.O. Wolfe lost one of his imported angels in a poker game and that this angel wound up on the grave of Hattie McCanless in the Old Fort Cemetery in Old Fort. And another of Wolfe's angels marks the final resting place of Fannie Clancy in the Bryson City Cemetery. There are those that claim one of these angels was more likely the source of

Thomas Wolfe's inspiration. So, go search out these stone angels, look into their eyes, try to read their expressions, and then turn away to look in the direction of their gaze. It's what Thomas Wolfe must have done. And they all are looking towards home.

Wolfe's Angel

A Cary Hash: Tiny Tank's Tick Tour

Author's Note: The Hash House Harriers arrived in North Carolina in the early 1980s with the formation of the Tar Heel Hash in Durham. Other hashing groups have formed over the years, including the Sir Walter Hash House Harriers of Wake County. Hashers get together on a regular basis to choose one of their kind to be a "hare." This hare will lay an intermittent trail marked with dollops of flour, and the remaining hashers — or "hounds" — will attempt to follow this trail while running, shouting misdirections at each other (it usually sounds like "on-on") and losing focus at the occasional mandatory pauses at strategically placed "beer checks." At the end of the hash there is always food and drink — the famous hash après. Hashers are known by their quirky yet somehow appropriate hash nicknames, names they earn after proving what they are made of over the course of several hashes.

It was a steamy hot July afternoon. I parked in front of the Tiny Tank abode in one of the many new subdivisions that have sprung up in Cary — that fabled bedroom community that sprawls over the Piedmont just west of Raleigh. The Tar Heel Hash House Harriers usually chase flour marked trails in Durham and Chapel Hill and leave the Wake County dollops of flour for the Sir Walter Hash House Harriers. But this hash would not be of the hide-bound type. At this hash, Tar Heel and Sir Walter hashers would commingle. The Tar Heel Hashers, consisting of me, Grumpy, 3Pints, Lickety Spit, Greg C., Micro, DooVarnay and the Bigfoot family, were present but severely outnumbered and outnamed by the Sir Walter contingent: Chickenman, Mrs. Chickenman (if a Tar Heel Hasher she might well be known as "Chickenwoman" but too bad for her she was a Sir Walter Hasher and so she was christened "Pullet"), Iceman, DuracElvis, Dicken's Cider (huh? is someone gonna explain that one to me?), Falsies, Southern Comfort, Pita, and Bigfoot's fleetfooted offspring whose hash name is the sound that Roy Orbison makes in the song "Pretty Woman" — yeah, that's right, the growling like sound indicating . . . well . . . appreciation. Henceforth we will indicate this hash name with the glyph "&." Sort of like Prince, or that is, the singer formerly known as Prince. So anyway, the hare, Tiny Tank (a high ranking officer in the Sir Walter Hash Harrier hierarchy, mind you) explained the Tar Heel Hashing rules to the Sir Walter crowd, warned us to beware the tick gauntlet he would run us through, supplied us with his home brewed tick-be-gone juice, and then took off with a bag of flour asking for a ten minute lead. Indolent hashers that we are, we gave him fifteen minutes and then shuffled off in pursuit — in no real hurry to confront the hungry denizens of Cary's tickworld.

 The streets we ran on in Cary had not been there long. Only a few years ago, this hash course would have been entirely in field and forest. Maybe that is why the ticks we met were numerous and surly, as Cary's expansion had crowded and concentrated them into the few remaining stretches of jungle TT had discovered for us to run through. There really was a surfeit of ticks — ready to adhere to our legs and then head north, despite the tick-be-gone juice smeared on our legs (which later turned out to be just sugar water anyway — very funny, Mr. Tiny Tank). And have you ever noticed that once you pick one tick off your leg, you keep feeling them all over your body even when they are not there? And all crawling north at that. Creepy. Well, we almost made it to Lake Crabtree Park but the Tanker instead diverted us through a series of construction sites of various Cary works-in-progress. We ran on huge bulldozer scraped fields, new roads with only the curbing in place, and some more forest trails. We had several Sir Walter style "hash halts" (not a bad idea, actually), a beer break (actually a great idea on this hot day in particular), and several opportunities for talk of ticks and tickchecks. At each stop Iceman attempted to groom various slow-reacting distaff hashers while continuously mumbling to himself about ticks and where they might be headed. When Iceman reached &, he found her not slow-reacting, and & soon rallied the troops, pried us away from the beer break and on to the home stretch and on-in. Someone said if we could beat & to the finish we'd get a beer. But we figured we'd get one anyway so we let & disappear in the distance and dogged it on home.

 The Tiny Tanker's après was quite nice despite the absence of that white grape beverage that we usually associate with Cary soirees — no brie either. The Tanker is still primarily a Sir Walter Hasher, after all — but he did have quite a selection of uncommon brewskis that he offered up to the thirsty, tick-covered horde. Iceman and DuracElvis complained that there was only one bag of Cheese Poofs (unlike the usual multi-bag Sir Walter après) but the naive Tar Heel Hashers were quite impressed with the spread. So we sat around on Tank's back porch, drinking his uncommon beer, scratching where the tick roamed, and speculating about the demise of Tank's running career after he becomes a family man in a few months time. About thirty minutes into the après, a sweaty tick-vehicle swept around the corner of the house and into our view. It was Rod & Staff, a Sir Walter hasher who had shown up, as is typical for the Sir Walter group, a half hour late and had run the hash course by himself. He was disappointed to have missed the tickchecks but amiable enough otherwise. Maybe these Sir Walter guys aren't so bad, I thought. So we sat. We talked. We drank more uncommon beer. One by one, all the Tar Heel Hashers left except me. Hey, these Sir Walter guys really aren't so bad — they got funnier. Tank got us some more uncommon beer. I got funnier. I started feeling kindly toward Cary and even its endless new subdivisions. I no longer perceived any problem with the Sir Walter Hash's Tank haring a Tar Heel Hash. I scratched my ankle and pulled a tick off my leg. I smiled kindly looking down at him. "Run free, little fella," I thought to myself and I flicked the little arachnid toward Iceman and DuracElvis. They looked up at me from the bag of Cheese Poofs. I smiled and saluted them with my beer bottle. They nodded back, with their own smiles. Ahhhh. The hashing life. Ya gotta love it! On-on!

For more information on hashing see:
- Tar Heel Hash House Harrier web page:
  http://pages.zdnet.com/commentateur/tarheelhashers
- Sir Walter Hash House Harrier web page:
  http://www.swh3.com/

Encounters with the Shad Fish and the Shad Bones

It's mid-March. The daffodils and the pear trees are abloom. The average daily temperatures are slowly rising. And the shad are running once again. After several years in the ocean, ranging as far north as Canada, the shad are entering the mouths of the Roanoke, the Chowan, the Cape Fear and the Neuse Rivers, and they are finning their way as far upriver as their bodies' stored energy can take them—assuming they aren't first stopped by dams, pollution or the determined shad fishermen who line the rivers in the early spring.

I know that today the shad are running in part because they always do in mid-March in North Carolina. I also know because I've seen the reports in the newspaper and on the internet of the increasing number of shad in the rivers. But the number one reason I know the shad are running is because I caught one today. The chase wasn't too difficult and I didn't get my feet wet. And I didn't use a shad dart. And it only cost me $1.99 a pound. Yes, I caught my shad in a local specialty grocery store, in the iced display case, surrounded by blue fish, salmon, rainbow trout and king mackerel filets. The trout was $8.99 a pound. But I was after shad—and I was undeterred by the little cardboard sign perched atop the three shad filets that stated: "Has lots of bones." Yeah, yeah, I knew that. I'd read the book on this famous American fish. Bones, schmones. And I had the recipe that was supposed to dissolve those ubiquitous shad bones.

The shad is a silvery fish built for speed. The mature females usually weigh 5-6 pounds and the mature males 2-3 pounds. Shad are found as far south as Florida and as far north as Labrador. Although originally an east coast fish, in 1871 some baby shad were introduced to the west and they flourished there. Like salmon, shad are born as far up freshwater rivers and streams as their parents can get before they spawn. The baby shad then float with the river waters down to the ocean and it is in the ocean where shad spend most of their adult life. The shad swim in schools, covering maybe two thousand miles or more in a year. It is believed that up to eighty percent of all adult East Coast shad spend a part of the summer in Canada's Bay of Fundy. But by summer's end they head back out to the open ocean and swim south. In early spring the shad return to the same rivers in which they were born and, leaving the salty ocean, they start upriver. They'll swim up their chosen river as far as they can—in some cases up to four hundred miles—until they are stopped by dams or until the water temperature reaches a certain point. Then they'll spawn, beginning the whole process again.

Fishermen who like a challenge like to fish for shad. Shad are fighters, with a lot of power and a lot of spunk. They jump out of the water a lot when on the line. I'm not a fisherman and I don't know fish, but it intrigues me that shad do not eat at all after they begin their spawning run and head upriver. During this time they can lose forty percent of their body weight. But it's during these spawning runs that fishermen catch the shad using their special shad "darts." The shad will strike these lures even though they are not eating anything. And, apparently, the shad never swallow the lures, they just hold them in their mouths. This combination makes the shad an exciting quarry for the scads of shad anglers who line the banks and shallows of some rivers when the shad begin their springtime runs.

Although the shad still run up east coast rivers in the twenty-first century, their numbers are greatly diminished from colonial times. The huge runs of shad in the 1700s and early 1800s were of immense importance to those living on the rivers. Celebrated author John McPhee published a book about the shad, calling it *The Founding Fish*. This was an important fish in American history. Thoreau empathized with the shad ("Poor shad! Where is thy redress?"). George Washington bragged about catching three hundred shad in his seine at Mount Vernon. Thomas Jefferson claimed there was no greater delicacy than the shad. And there is a much repeated, though probably apocryphal, story of how an early shad run on the Schuylkill River in February or March of 1778 saved the Continental Army camped at Valley Forge from starvation. Americans and America relied on these huge schools of shad surging up its rivers every spring.

It was no different in North Carolina. The *WPA Guide to the Old North State* tells us that even as far inland as the site of Trading Ford (near present-day Salisbury): "In Colonial times settlers annually met the Indians to trade, especially for shad, near here on the Yadkin River." And when, as a young man, Daniel Boone lived in the Forks of the Yadkin, he would catch shad with a seine "at the Shoals just above Dutchman's Creek." Toward the end of his long journey in the Carolinas in 1701, John Lawson described the following encounter: "We were forced to march, this day, for Want of Provisions. About 10 a Clock, we met an Indian that had got a parcel of Shad-Fish ready Barbaku'd. We bought 24 of them, for a dress'd Doe-Skin and so went on, through many Swamps, finding, this day, the long ragged Moss on the Trees, which we had not seen for above 600 Miles." Later in his account, Lawson noted that "[s]hads are a sweet Fish, but very bony; they are very plentiful at some Seasons." The Native Americans had been experts in catching shad long before the colonists arrived. In his book *The American Indian in North Carolina*, Douglas Rights, stated: "The Indian was an expert fisherman. He constructed weirs for copious catches. In the rivers two rows of stone were placed to form a V-shaped figure with the point of the V downstream where the weir or trap was placed. One double fall, shaped like a W, is in the Yadkin River south of Salisbury. Th fishermen beat the water above the fish-fall and drove the fish into the traps. In the spring great catches of shad were made in the rivers. Many of these stone fish-falls may still be traced in the streams, since the later residents in this region have kept them in repair." The shad is, thus, an historic fish in these parts. Why not experience this sweet, tasty, bony fish just like so many did centuries ago?

So back at home I unwrap my three shad filets and look at them. Little bones are sticking out everywhere. I plan on slowly baking this fish following one of the recipes in McPhee's book. On a whim I pull out my old copy of the *Joy of Cooking* cookbook and look up shad. It's there, and it says: "Whole fish are often slow-cooked in an attempt (usually futile) to dissolve bones." Well, okay. I guess we will see who is right about this bone dissolving business. The shad go in the oven, covered with foil, and I vow to not even open the oven door for several hours so those bones will have little choice but to dissolve away.

Hours later I serve dinner. So was the shad tasty? Indeed it was. It was sweet and I can understand how Thomas Jefferson could have rhapsodized over its flavor. While eating my shad, did I share a sense of identity with all those Native Americans, colonists and citizens of the United States who caught and consumed all those shad fish in the eighteenth and nineteenth centuries? I most certainly did. Only if I'd caught the shad in my own seine would the experience have been more equivalent. Did the bones dissolve? Not a one. That is the boniest fish I've ever eaten. It is more work creating a forkful of deboned shad than it is getting a forkful of crabmeat from a boiled blue crab. But ultimately all the shad (and a few shad bones ... a few hundred or so) were consumed—though it took several days. And I'm sure the occasional pain I get in my stomach is another of those undigested shad bones making its slow way south. Lawson had warned me—though sweet, shad are simply a very bony fish.

There remained, however, one item left to do to complete my shad experience. So a couple of weeks later I journeyed eastward to Grifton, North Carolina, where on the banks of the shad filled Contentnea Creek, the Annual Grifton Shad Festival was in full swing. The Festival has the usual crafts fair, rides, cloggers, etc., but it also features free fried shad, a shad fishing competition, a fly rod casting competition, and my personal favorite, the shad toss. As I wandered through the Festival, over to the shad toss, I munched on some of the free fried shad I had waited in line for. Pretty tasty, I had to admit, but still no obvious bone dissolving. The free fried shad was going fast but there weren't many in line for second helpings.

At the shad toss, from out of a cooler full of ice and shad I picked up and held my first intact shad fish. It was sleek, silvery and a pretty good looking fish. Smooth and somewhat bullet-shaped, it looked like it would have been a fast swimmer. It looked bony too. I hefted it a couple of times then walked over to the shad toss official. He motioned me forward and I adjusted my grasp around the shad's slimy midsection. Three steps forward and I flung that fish toward eternity. High up against the blue eastern Carolina sky, flashing in the sunlight, the shad's tail flipped back and forth as though it were swimming hard through the air. Its downward arc brought it bouncing and skimming off the grass to where it finally lay still on the turf. They marked my throw and handed me back my shad. As I walked back to the cooler another silvery shad flew high through the air, mouth open and tail flipping. I turned away from the ice cooler with my shad still in hand. I walked back to the shad toss official. I'd have to throw again.

The Shad Toss at Grifton's Shad Festival

Barbecue Church, Cross Creek, and Flora Mac

I drive south from Sanford on Highway 87 into Harnett County and exit on Highway 27 which my map says is also called Barbecue Church Road. After a few miles I see the historical marker for the Barbecue Church, "founded in 1757 by Scottish Highlanders." The present Barbecue Church, built about 1895, stands about 100 yards away on my left.

Behind the church building is a cemetery. I walk through the headstones toward the tree line and soon find the oldest section of the cemetery—near where the original church building must have stood. Sure enough, in the woods at the edge of the cemetery is a large stone "Cairn of Remembrance" marking the site of the original log church, which a plaque indicates was "the first permanent place of worship in Harnett County." I walk further into the woods and find a leaf covered path leading down a bluff toward the bottom land around a creek. At the end of the path, at the base of the bluff, I find the inverted pipe that now protects the Barbecue Church spring—the spring that quenched the thirst of the Scottish Highlanders that had migrated to the area and founded the church in the mid-1700s—the spring that quenched the thirst of one famous church member in particular—Flora MacDonald, the heroine of Bonnie Prince Charlie's successful escape from the English dragnet in 1746. It is said that she was the most famous woman in pre-Revolutionary War North Carolina. I linger at the spring, listening to the rustle of the leaves. After a time, I head back up the path. Next stop is Cross Creek

Charles Edward Stuart, aka Bonnie Prince Charlie, grew up Catholic in Rome, Italy, but believed he was rightfully destined to be the king of England. With encouragement from France, Bonnie Prince Charlie, at the age of 25, landed in Scotland and gathered his Scottish supporters for an attack on the English. On 16 April 1746, at the Battle of Culloden on Drummossie Moor, the English forces, led by King George's son, William, the Duke of Cumberland, decisively defeated Prince Charlie's army of Scottish Highlanders. The bonnie prince escaped from the field of battle and headed north into the Highlands. The Duke of Cumberland gave chase and devoted his large naval and land forces to the capture of the prince. Cumberland also engaged in brutal reprisals on the Scottish people, earning the nickname the "Butcher." A reward was placed on the prince's head and those suspected of aiding the prince's escape were jailed and had their property confiscated. Nevertheless, Bonnie Prince Charlie eluded capture for five months and somehow was able on 20 September 1746 to board a French ship sent to save him and to sail through the English blockade to safety on the continent.

Prince Charlie's five month flight to escape the English manhunt is a tale full of close calls, bravery and luck. Despite the punishments meted out by Cumberland's men to those who aided the prince, Scottish Highlanders continued to help the prince as he hid in sheds and caves, trekked across mountain and valley, and traveled in small boats to various islands thought to be safe. One famous chapter in the escape involved a comely 24 year old Scottish lass who risked her family's and her own health and wealth by smuggling the prince to the Isle of Skye.

Flora MacDonald, two years younger than the bonnie prince himself, was asked to help with his escape and by all accounts she never hesitated to agree. The prince disguised himself in women's clothes and became "Betty Burke," a servant of Flora. Their small band passed through several checkpoints in this disguise, succeeding largely through Flora's calm, charm and believability. Betty Burke's appearance and awkward gait were explained by noting that she was an Irish girl. Flora accompanied the prince on the long boat ride to the Isle of Skye during which they were fired upon and endured a sudden squall that came near to swamping their small skiff. But the escape was successful and after eleven days together, Bonnie Prince Charlie parted from Flora on 1 July 1746 to continue his flight. It is said that at their parting, the prince kissed Flora's hand and told her: "For all that has happened I hope, Madam, we shall meet in St. James' yet." But Prince Charlie never made it to London, and also never communicated with Flora again. Denied another chance to recover his crown, Prince Charlie died in exile in 1788, reportedly a bitter, cantankerous alcoholic.

But in July 1746, Flora MacDonald's troubles had just started. English authorities discovered her role in aiding Prince Charlie's escape and she was arrested. In November of 1746 she was taken to

London and imprisoned in the Tower of London pending her trial. She remained in custody for 12 months. But, as sometimes happens, Flora MacDonald's story, her obvious courage, capabilities and beauty, captured the imaginations of the public and the aristocracy. Her case never went to trial, and instead she was released and, for a time, became "an idol of London society." Flora was visited and feted by London's upper crust. Many artists asked her to sit for portraits. Dr. Samuel Johnson, upon meeting her, described Flora as follows: "[Her name] will be mentioned in history, and if courage and fidelity be virtues, mentioned with honor. She is a woman of middle stature, soft features, gentle manners, and elegant presence." Flora was an object of great honor and celebration upon her return to the Scottish Highlands. In 1750, she married Allan MacDonald, a landed gentleman.

The MacDonalds' life must have been idyllic for a time. But it is believed that financial troubles gradually grew for Flora and Allan MacDonald so that in 1774 they decided it best to immigrate to America with five of their seven children. After the Battle of Culloden many Scottish Highlanders had immigrated to America and many had settled in North Carolina in the area around present day Fayetteville. That is where the MacDonalds headed. In 1774 the community was called Cross Creek.

According to the *WPA Guide*, in the 1740s, "a group of expatriated Scots, men who had escaped 'the penalty of death to one of every 20 survivors of Culloden,' established a gristmill and village at Cross Creek ..., where they found two streams crossing each other." The *Guide* also states that the two streams "met and apparently separated, forming an island of some size. It was said that the streams, when swollen from the rains, actually crossed each other in their rapid course." But, alas, this fascinating crossing was eliminated by the construction of a cotton mill on the site by a Frenchman named De Gross in the 1740s. But the name stuck. Allan and Flora settled in a home on the banks of Cross Creek in 1775.

Fayetteville has grown up around and over the old settlement of Cross Creek. The Flora MacDonald house is long gone but the house site is said to lie on the northeast corner of the intersection of Green and Bow Streets—close to the traffic circle at Market Square.

Sure enough, at this intersection the meandering Cross Creek flows under the highway past a small park and a First Union Bank where the house must have once stood. A small map in the *WPA Guide* shows the old site of the crossing creeks nearby, on the north side of Rowan Street. But the present day land no longer fits this old map. It looks like at some point Cross Creek was diverted into a new stream bed so that it now joins Blounts Creek near the Hawley Lane bridge on the south side of Rowan Street rather than the north side. And the old stream bed of Cross Creek, which used to form the western border of the Cross Creek Cemetery, and the actual site where Cross and Blounts Creeks may have once actually crossed each other, appears to have been buried deep under many loads of backfill. A city maintenance facility enclosed by a chain link fence now borders the cemetery sitting atop this landfill—high above the old Blounts Creek ravine, which now carries the waters of both Blounts and Cross Creeks. The sites are there but it takes some imagination to see them as Flora MacDonald would have seen them in 1775.

The MacDonalds' time in America was not destined to be easy or lacking in ironic twists. First of all, the MacDonalds had settled in Cumberland County, formed in 1754 and named after the bloody Duke of Cumberland—the English victor at the Battle of Culloden, the pursuer of Bonnie Prince Charlie and the man whose savage reprisals against the Scottish Highlanders after Culloden was the cause of the large Highlander migration to America in the 1740s and 1750s. It was also true that most of the Highland Scots who left Scotland for America following the Battle of Culloden were forced to solemnly pledge allegiance to the British king, promising that if they broke their pledge they would be "cursed" in all their "undertakings, family and property." This oath presented a problem in the late 1770s for the Highlanders in North Carolina when their friends and neighbors sought their support for the burgeoning movement for American independence from Britain. The Highlanders had good reason to detest the British monarchy and to feel great sympathy for the American revolutionaries—but whether because of the oath or because of their long history of being ruled by kings, the Highlanders generally chose to remain loyal to the Crown. Flora and Allan MacDonald were not exceptions.

When the royal governor of North Carolina sought to raise troops to oppose the American revolutionaries, he turned to the Scottish Highlanders in the Cross Creek area. Many of these men were military veterans who had fought at Culloden. They were willing recruits for the loyalist side. One of their leaders was Flora's husband, Major Allan MacDonald. In February of 1776 the Highlander army gathered at Cross Creek and prepared to march toward Wilmington to engage the rebel patriots. There to exhort and inspire the soldiers was the most famous woman in North Carolina—Flora MacDonald. Mounted on her white pony, Flora spoke to the troops in Gaelic, rallying them to the loyalist cause by speaking of Highland heroism, devotion and honor. The troops responded enthusiastically and began their march. As they passed by her, it is said that Flora MacDonald called out to each clan its Gaelic battle-cry. The site of Flora's review of the troops as they marched to battle is where Cool Spring Street spans Cross Creek—on the south side of the creek. The spot is marked by an historical marker.

The Highlander army, started out with glory, but met a vainglorious end. Nine days after leaving Cross Creek, on 27 February 1776, in the early morning darkness, the Highlanders sought to cross over Widow Moore's Creek in Pender County. The Patriots under General James Moore had, however, been expecting them, and had removed planks from the bridge and greased the logs that remained. When the Highlanders surged forward, shouting "King George and broadswords!" the Patriots responded with withering rifle fire. The Patriots were well entrenched on their side of the creek, and their rifles decimated the charging loyalists. The Highlanders lost about seventy killed or wounded while the Patriots had none killed and only two wounded. Eight hundred and fifty Highlanders were taken prisoner. The Patriots' victory at the Battle of Moore's Creek Bridge was decisive, and eliminated the Highlanders as a significant factor in the fighting.

The defeat profoundly impacted Flora MacDonald. Her husband was imprisoned and taken north—not to be released until 1778 as part of a prisoner exchange. Flora remained in North Carolina but was the object of many petty and not-so-petty reprisals by those who resented her role in supporting the loyalist cause. Her estate was confiscated by the Act of November, 1777, passed by the Provincial Congress sitting in New Bern. Her grown daughters were also mistreated by members of the Patriot army. Flora resolved to leave America and return to Scotland. She journeyed to Wilmington where she was forced to sell what remained of her silver service (some of which had been gifts from wealthy London admirers) to raise cash for her trip home. Some of these pieces are still in the possession of families in the eastern part of North Carolina.

Finally, in 1779, after five years in America, Flora, with only one of her daughters, set sail for Scotland—leaving from Charleston, South Carolina. Her travails were not quite over though. Her ship was attacked by a French vessel, and, it is said, that she stayed on deck during the hand to hand fighting—urging the British seamen on and suffering a broken arm during the melee. Flora returned to the house she and her husband had left in 1774, where her husband, Allan, was not able to rejoin her until 1783. Flora MacDonald died in 1790 and is buried on the Isle of Skye. It is said that she was buried shrouded by a sheet on which Bonnie Prince Charlie had slept during his flight in the Highlands in 1746. She had kept this sheet, and a lock of the prince's hair, for forty-four years—she took them with her to America and they were one of the few possessions with which she was able to return to Scotland. A monument to Flora MacDonald on Castle Hill, Inverness, Scotland, has the following inscription: "As long as a flower grows in field the fame of the gentle lady shall endure."

I leave the site of the Highlanders' "March-Out" on Cool Spring Street, cross over Cross Creek and drive by the Cross Creek Cemetery on my right. I turn left on Rowan Street and then right on Ramsey Street heading north. I'm still thinking about Flora MacDonald's difficult five year stay in North Carolina—and just what those early settlers could have seen that made them think that two creeks actually crossed each other. I almost miss my left turn into the Veteran's Administration Hospital where, underneath a line of trees on the south side of the entranceway, running from the Hospital back to Ramsey Street, is a well preserved segment of a Confederate trench—built to defend the city of Fayetteville from federal attack in 1865. I park and walk along the earthworks. But my heart is not in this trench. I'm still thinking about the meandering course of Cross

Creek as it must have been in the mid-1700s as it flowed toward its storied meeting with the waters of Blounts Creek. And I'm thinking about the strange turns in the life of Flora Mac and her short stay on the banks of Cross Creek.

Flora MacDonald marker in Fayetteville

Into The Waccamaw

The air hanging over Lake Waccamaw is absolutely still. No movement ripples the water. The lake's surface is a perfect reflection of the January sky. The air temperature is 46 degrees. I see some sort of water bird silently flapping over the water near the southern shore. It disappears into the forest at the lake's edge. There is no sound at all. I stand on the shore and gaze off into the distance. For a time. It dawns on me. I'd like to make a wake in that clear still water sometime. To swim across the lake. As I walk back to my car, I think to myself that I'll write a letter.

18 January 2002

Lake Waccamaw State Park
1866 State Park Drive
Lake Waccamaw, N.C. 28450

Dear Sir or Madam,

I am a distance swimmer and I am interested in swimming across Lake Waccamaw. I believe there used to be an organized swim across Lake Waccamaw but it may have been discontinued several years ago.
Can you tell me if there is any organized swim across Lake Waccamaw, either scheduled or to be planned? If not, would a swim across Lake Waccamaw be allowed? I would expect to have several other swimmers and attendants in a boat accompany me on the swim, and to complete the swim responsibly and safely.
If you would allow but discourage the swim I would appreciate knowing of your concerns. Thanks for your help and advice.

Sincerely,
Tom Fowler

January 25, 2002: No response yet from the Waccamaw State Parksters. Didn't expect any so soon anyway. But I did swim a mile and a half today in the pool. Waccamaw is like a four or five mile swim so I do need to get in some serious miles. Maybe thirty miles in January?

Columbus County's Lake Waccamaw is five miles long and three miles wide—and shallow. It's about eleven feet down at its deepest, and it averages maybe seven feet deep. Waccamaw is a spectacular example of a Carolina Bay. Carolina Bays are elliptical depressions in the earth spread out all along the east coast from Maryland to Georgia, although most are found in the Carolinas. There are thousands of them (some say half a million). Many are small and many are now completely filled with vegetation. But some are still shallow lakes. In North Carolina's Bladen County, Jones and Salters Lakes in Jones Lake State Park, and White, Bay Tree and Singletary Lakes in Bladen Lake State Forest, are also good examples of Carolina Bays.

February 5, 2002: All right. I have a letter from Lake Waccamaw State Park in my hands. If it isn't an outright refusal, I bet it will discourage my proposed swim. I'll bet. Maybe I'll just let the letter sit awhile before I open it. Then I open it.

Dear Mr. Fowler:

Thank you for your letter concerning your interest in Lake Waccamaw and swimming. Lake Waccamaw is very popular for swimming and other water sports. Swimming to any extent in the Lake is allowed all year (weather is your option) round. It is very highly recommended that you have other swimmers accompany you any time that you are in the water and having close access to a safe watercraft while you are in the water is also a good precaution.

There is an annual organized swim across Lake Waccamaw. It is known as the "Labor Day Swim", as it is held on Labor Day of each year. Please contact Mr. John McNeill and he will be able to give you the details about participating in the event. His address is John McNeill, Sr., P.O. Box 339, Whiteville, N.C. 28472

If you have any questions, please let us know. Thank you and have a good day.

Sincerely,
Janice Mercer, Office Assistant
Lake Waccamaw State Park"

Excellent! Interestingly, though, the Lake Waccamaw State Park stationery that this letter is written on has, at the bottom right hand corner, an image of a smiling, green alligator. This is why it is highly recommended to swim in groups, huh?

5 February 2002
John McNeill, Sr.
P.O. Box 339
Whiteville, N.C. 28472

Dear Mr. McNeill,

I understand that you may be able to tell me about the "Labor Day Swim" across Lake Waccamaw. I'm a distance swimmer and am interested in participating in such a swim. Could you send to me the details about the event and how to enter? I've enclosed a self-addressed stamped envelope.

Thanks very much for your help and information.

Sincerely,
Tom Fowler

Theories abound as to how the Carolina Bays came to be. The most popular and oft-repeated theory is that a meteor shower visited the area during the late Pleistocene Era and each meteor striking the earth dug a comparably sized oval depression in the ground. This explanation is consistent with the fact that the elliptical Carolina Bays all seem to be oriented along the same northwest to southeast axis. But no meteorite fragments have ever been found in or near the Bays. So maybe the alternative "wind and wave" theory is right after all.

On February 13, I received this letter from John A. McNeill, Sr.:

> *Thanks for your letter of Feb. 5. I am eager to respond. The Lake Waccamaw Labor Day Swim is a very informal event which honors Judge Lee J. Greer, a devoted distance swimmer, who taught hundreds of people to swim, and was an inspiring leader in Columbus County and North Carolina. The event is held every Labor Day at Lake Waccamaw. There is no registration other than being placed on the long list of participants. Just show up on the beach in front of Boys Home (at 8 a.m.). It is not a race. It is a personal achievement. A medal, the Lee J. Greer award, is given to each swimmer as he finishes the swim. The course is decided each time so that winds are favorable. The distance is about 3.6 miles. Each swimmer is required to have a boat and observer alongside. Course markers and Rescue Squads are there to assist. Hope to see you on Labor Day.*
>
> *John*

Okay! Count me in. Seven months to train and only 3.6 miles to swim—and there will be plenty of other splashing swimmers to scare off the Waccamaw alligators. Guess I'll head off to the pool for a couple of laps.

The Carolina Bays all have clear, shallow waters, and are generally surrounded by pocosins—the encroaching peat bog forests that gradually fill in the depressions with vegetation—with a sandy shoreline at the southeastern end of the lake and a raised bluff on the northwestern edge. The Carolina Bays generally get their water from the rains rather than from inflowing streams—although Waccamaw receives some inflow from Big Creek on its northern end. The water in the Bays is generally acidic, limiting the fish populations. But

Waccamaw, because of a limestone intrusion in its waters, is less acidic and so supports a wide variety of fish. Some, like the Waccamaw darter, the Waccamaw silverside, and the Waccamaw killifish, are found only in Lake Waccamaw.

June 3, 2002: Training is going well. I swam forty-two miles in May, and swam a two-mile lake swim in Virginia last weekend—scared those fishes for over an hour.

Artifacts from Waccamaw-Siouan tribes have been found in the Waccamaw area indicating Native American habitation for several thousand years. Famed naturalist John Bartram, sometimes called the "father of American Botany," visited Lake Waccamaw in 1734 and called it "the pleasantist place" he had ever seen in his life. The first white settler on the lake was John Powell who settled there in 1745. Although some say he was born in Georgia and others in Alabama, local legend has it that Osceola, the famous chief of the Seminoles during the Seminole Wars in the 1830s, was born on Waccamaw's shores—and that Powell might have been his father. In his youth, Osceola was known as "Billy Powell."

August 3, 2002: Forty-eight miles in July! My goggles don't leak, I've got my sons lined up to man the safety boat, I'm all set. Time to ease off on the high mileage.

The lake is drained by the Waccamaw River which flows south through Green Swamp then into South Carolina where its black waters flow finally into the Atlantic. In 1895 a steamer sailed up the Waccamaw River from Georgetown, South Carolina, to enter Lake Waccamaw. Dams built on the river subsequently limited its navigability. In the early twentieth century Green Swamp was lumbered to produce cypress shingles which were shipped by boat across Lake Waccamaw for mule transport to the nearest train station. In 1976 the Lake Waccamaw State Park was formed when a 273-acre tract of land was purchased by the N.C. Division of Parks and Recreation. The Park is now about 1,732 acres.

October 5, 2002: I did enjoy Labor Day 2002 but it wasn't exactly what I expected. I swam not a stroke. Didn't even get my big toe wet. My goggles never even had a chance to leak. No, sometimes the course of human events, dictates that "Tom, you have some inescapable familial duties that you must attend to and Columbus County and that big Carolina Bay, however appealing, are just too far away to let you get there and back and still do what must be done." And I couldn't disagree. The priority list was a no-brainer. So I spent a fulfilling, productive Labor Day without burying my face in murky waters and splashing around for several hours. Okay. But today, it's warm and clear—the first Saturday in October. And I happen to find myself standing beside the dock at Lake Waccamaw State Park. My bare feet are sunk in the muck at the edge of the big Carolina Bay and I'm looking northwest toward the far shore—maybe five miles away. My no-leak goggles are already strapped on. My multi-colored Speedo catches the rays of the morning sun. I stride purposefully forward. The water reaches my knees. My waist. I lean forward and push off. I'm borne by the still waters. Into the Waccamaw.

Lake Waccamaw State Park, 1866 State Park Drive, Lake Waccamaw, NC 28450 (telephone: (910) 646-4748), is located in Columbus County, 38 miles west of Wilmington and 12 miles east of Whiteville. The park is located on Martin Road off SR 1947. Look for signs on US 74/76 and NC 214.

Lake Waccamaw

Running the Scenic Crest of the Sauratown Mountains

If you are driving south on North Carolina's Highway 77, right where the road begins the drop down the face of the Blue Ridge, and you gaze off to the east on a clear day, you can see three distinctive mountains rising out of the Piedmont on an east-west alignment. These are the Sauratown Mountains. They are not a part of the Blue Ridge, and are not much over 2,500 feet in height—but they are prominent nonetheless because they loom 1,500 feet or so above the surrounding flatlands. Pilot Mountain is the closest, about thirty miles away. The Saura Indians called it "Jomeokee," the great guide or pilot. Because of its distinctive rock cap summit, Pilot Mountain has served as a landmark for countless travelers for many centuries. Hanging Rock is the mountain farthest east and Sauratown Mountain is the one in between. If you are in a hurry or on business you keep driving past Mt. Airy and on to Winston-Salem. But if you are of an adventuresome bent, you can't help but wonder what it would be like to run or hike from the top of Pilot Mountain over Sauratown Mountain and finally stop at the summit of Hanging Rock. From the top of the Blue Ridge, it doesn't look to be too far of a trek.

It's August 26. 8:30 in the morning. The Big Pinnacle, the rock cap which sits atop Pilot Mountain, is awash in mist. I'm on the trail which encircles the Big Pinnacle, leaning casually against the exposed rock. I'm ready to run. My road crew synchronizes her watch and nods at my exhortations concerning the great photo opportunity before her. Me with Pilot Mountain in the background, of course. I glance over my shoulder and peer through the mist eastward toward Hanging Rock. Can't see it—too foggy. At the signal, I push off from the Pinnacle and slouch down the rocky trail. In only about 100 yards I reach the parking lot near the summit. My road crew has already fired up the sag wagon and taken off. I jog the curving, banked roadway 2.5 miles down hill to the entrance to Pilot Mountain State Park.

From here I will be following the Sauratown Trail, along some roads and trails on private and public lands, over Sauratown Mountain and ultimately to Hanging Rock State Park. I have my Sauratown Trail map and directions in a plastic baggie and I study both items as

I idle down a deserted country road. I wonder vaguely about country folk hereabouts and whether their friendly doggies will be tied up or roaming. Guess I'll find out. Despite numerous turns on little back roads, the directions are good and I don't lose the trail for another three miles. But then I take an educated guess and turn right when I should have gone left. I jog about a mile and find myself in the little town of Pinnacle—which is not where I'm supposed to be. No sag wagon in sight so there is nothing for it but to retrace my route. I pass the point of my wrong turn and soon bump into an actual trail in the woods and, miraculously as it seems to me, a sign that says "Sauratown Trail." Must be the place, eh? The trail does turn out to be real—narrow and rarely traveled, yes, but still I never get lost again after I get on this trail.

The main problem with this trail, however, is that apparently no one has hiked it all summer. Every twenty feet or so there are huge sticky, gooey spider webs spanning the trail. I run through every one. If you do the math you will realize that there is a huge sticky wad of spider webs accumulating on my head and chest as I run over this trail of several miles, not to mention the annoyed spiders that occasionally come along for the ride. I do my best to perspire profusely to wash away the slimy webs. I ford the West Prong of the Little Yadkin River getting my feet wet. I emerge on a road just minutes before my loyal road crew was going to give up on me—or so she said. Energized by food and drink, I enter the woods on the other side of the road and head up the steep trail on the side of Sauratown Mountain. Half way up I realize I am a might beat. Wet shoes, spider webs, trail still headed up—I am over half-way aren't I? The map isn't really very clear on this point. But the sag wagon has moved on so it's no use turning back.

At the top of Sauratown Mountain the trail ends and I'm back on a dirt road. My wonderful road crew is looking ever more beautiful every time I see her and her car stopped by the side of the road. But I can't tarry. The long dirt road down Sauratown Mountain has the worst dogs. But mostly they just hurry out to investigate, scare the bejesus out of me, make faces and a lot of noise. The route winds around on a few more roads and then turns back into the woods and I pass a sign that says I have entered Hanging Rock State Park. All right! That's a relief because I am getting busted. But I'm not just about there. It is a long stretch on the park trails, Huckleberry Ridge and Moore's Knob, until I reach the lake and bathhouse at Hanging Rock (but at least these trails had been recently hiked so no more spider webs across the trial!). The drop-dead gorgeous road crew is there in the parking lot with a syrupy soft drink and several kind words. Excellent drink. Excellent woman. She looks at me pityingly. I'm guessing I don't look so good. She says that I don't really have to climb the rest of the way to the top of Hanging Rock. But we both know that, actually, I do. So we walk to the last trail head and start up. It isn't so much further. When I scramble over the rocky outcrop and reach the very highest point of the Rock that Hangs, the official road crew timepiece pronounces: 5 hours 18 minutes and 5 seconds. A time to beat for all you serious adventurers out there. But if you try it, don't miss that left turn near Pinnacle. And on the trails run behind someone taller than you if you can. It helps with those spider webs.

More information:

Pilot Mountain State Park, Route 3, Box 21, Pinnacle, NC 27043; Telephone: office (336) 325-2355; fax (336) 325-2751; Directions: Located twenty-five miles northwest of Winston-Salem on US 52. Hanging Rock State Park, PO Box 278, Danbury NC 27016; Telephone: office (336) 593-8480; fax (336) 593-9166; Directions: Located in Stokes County, four miles northwest of Danbury. The park entrance is on Moore's Spring Road (SR 1001), which lies between NC 8/89 east of the park and NC 66 west of the park. For a map of each state park and its trails go to *http://www.ils.unc.edu/parkproject/Maps.html* and click on the appropriate map. For the Sauratown Trail, see Allen De Hart's *North Carolina Hiking Trails*, 2nd Edition, 1988, at pages 438-440.

Pilot Mountain, with Sauratown Mountain in background

Burnt Churches in South Carolina's Lowcountry

In the early 1700s, the lowcountry of South Carolina was quick to attract European planters, who would become wealthy growing indigo and rice, with the crucial help of the forced labor of a substantial population of African slaves. The second oldest town in South Carolina, Beaufort, was founded in 1711, and in the 1740s the local planters built a church, fifteen miles northwest of Beaufort, that was said to rival the finest churches in Charleston. The Church of Prince William's Parish had Greek temple columns and three and a half foot thick brick walls. The church must have been an important symbol for the lowcountry aristocrats around Beaufort — and that may have led to its visitation by British General Augustine Prevost during the Revolutionary War. In 1779, the British Army under General Prevost invaded the area and burned the church — though much of the walls and columns refused to fall. It is said that for a time parishioners continued to worship in the ruins of the church.

In the 1820s the church was rebuilt and renamed the Sheldon Church of Prince William's Parish — named either for the nearby Sheldon Hall Plantation or for Gilbert Sheldon, the seventy-eighth archbishop of Canterbury. Legend tells us that sometime after the church's second consecration, a Beaufort woman delivered to the church the parish's original communion silver which had been thought lost in the 1779 fire. The woman explained that a British soldier had saved them from the fire and given the silver to her for safekeeping. But the church would once again prove an unsafe haven. Sheldon Church was burned once again in 1865 by the Union Army under General Sherman as it marched north toward Columbia and Raleigh intentionally destroying much of what it encountered. And once again much of the church refused to collapse. Much of the brick walls and columns remained standing, and remain to this day. The ruins of the church are still used every year by descendants of the original parishioners to worship on the second Sunday after Easter.

The ruins of Old Sheldon Church are reached by driving north on Highway 21 from Beaufort. Turn left or west on Highway 17, and after about a quarter mile, north on Old Sheldon Church Road. The ruins are just down this road.

Nearby, east of Beaufort, is St Helena Island, the site of another, much smaller, burnt church. St. Helena Episcopal Chapel of Ease was built in the 1740s for the island's wealthy planters. The chapel was built of tabby — a type of cement made from oyster shells, lime and sand. But in 1886 when the fire locals were using to clear land burned out of control, the chapel was destroyed — although much of the external structure remains standing. It is a quiet, peaceful and calming spot. To get there, head east from Beaufort on Highway 21 and turn right or south onto Land's End Road. Down the road on your left you will see the ruins of the chapel.

A third burnt church is found further inland, about ten miles southeast of Walterboro in Colleton County. It was the first church established in St. Bartholomew's Parish and the original wooden building, built in 1726, was burned down by Yemassee Indians in the early 1750s. A brick chapel, Pon Pon Chapel of Ease, was built on the site in 1754. Pon Pon was the name of the Indian village that formerly occupied the site of present day Jacksonboro. The Pon Pon Chapel was ravaged by fire on several more occasions. By some accounts, it burned during the Revolutionary War, and others say it burned in 1801 and again in 1832. The ruins also suffered damage in 1959 when Hurricane Gracie struck. At some point in the early 1800s the chapel acquired the sobriquet, the Burnt Church — and that is how it is commonly referred to today. The chapel was located on the old stagecoach road connecting Charleston and Savannah — the same route that George Washington would follow on his Southern Tour in 1791. The ruins of the Burnt Church are found on Parker's Ferry Road just to the northwest of Jacksonboro, in Colleton County.

The ruins of Old Sheldon Church

Stalking the Elusive Carolina Petroglyph

I'm driving north. U.S. 301. I pass through Enfield and head toward Halifax. I'm looking for the bridge over Fishing Creek. I'm hunting—hunting something that might be found lounging on the banks of the creek—or maybe it's long gone. That's why it's a hunt. What I'm tracking down is something not mentioned in recent books on tourist attractions in North Carolina, or North Carolina historic sites. There won't be any visitor center or historical marker to aid my search. But if it is there, I should find it. I'm hunting a petroglyph—or at least what I think must be a petroglyph. I'm guided entirely by a single entry in the *WPA Guide to the Old North State* published in 1939. I've seen no other reference to this petroglyph in the literature. But the *WPA Guide* states: "U.S. 301 crosses Fishing Creek near which bones of an ichthyosaurus were excavated some years ago. On the creek bank is a large flat stone impressed with human and animal footprints and intricate designs." I'm hoping to park next to the bridge, walk down to the creek, where I'll spot that flat stone. And then I'll marvel over the ancient carved footprints and intricate designs.

Petroglyphs are inscriptions or carvings on naturally occurring rock faces. Native Americans have left thousands of these carvings across the United States. Petroglyphs sometimes clearly portray animals or hunting scenes, but many carvings are of hand prints, foot prints, and bizarre human like figures that are not easily interpreted. Pictographs are paintings on naturally occurring rock faces. Many petroglyphs and pictographs are of a great variety of geometric symbols or figures. Some petroglyphs were made by those of European heritage in more recent times. But regardless of their origin, it is true that these rock carvings and paintings are not fast moving. You'd think the petroglyph hunter would have the advantage. But you might be wrong.

The existence and location of petroglyphs and pictographs are often well protected secrets by those who know of them. That's why directions to the rock art sites, or even mention of their very existence, sometimes disappear from the literature. Unfortunately suppression of this information has some justification. Vandalism of well known petroglyph/pictograph sites seems to be as common as it is inexplicable. There is a lovely, unmarked pictograph site I visited in southern California (with an actual shaman's cave) where every rock face is covered with spray painted graffiti. And then there is Paint Rock—a sheer 100 foot cliff overlooking the French Broad River near Hot Springs, North Carolina. According to the *WPA Guide*: "The 1799 boundary commission, surveying the boundary line (NC and Tennessee) reported that the red 'stains' on the rock resembled the figures of 'some humans, wild beasts, etc.'" Though the *WPA Guide* suggests these figures are just oxidation of iron in the rock, it seems fairly clear that these were Native American pictographs. I couldn't find them anymore when I visited Paint Rock. But I did find a sheer rock cliff completely defaced with spray paint.

On a recent visit to Wyoming, my sons and I stopped by Register Cliff—another sheer rock face that rises in a pretty valley next to the famed Oregon Trail near the town of Guernsey. Register Cliff is covered with carved names and dates, ranging from those who traveled the Oregon Trail in covered wagons in the mid-1800s to teen-age tourists to the rock who had too much time on their hands in the 1980s and 1990s. Carvings from the 1880s overlay carvings from the 1840s and 1850s. And 1940s and 1950s inscriptions are scratched in between—with more recent spray painted versions over all. I guess it is all graffiti, whether 150 years or 15 years old. In time maybe even the 1960s or 1970s graffiti may become more interesting. But now it's just sad and annoying.

Okay, I see the bridge up ahead and the sign says "Fishing Creek." Here we go. I pull over, park and hike to the river. The water is high. I scan both banks. I walk to the downriver side of the bridge and again scan both banks. I walk a short ways along both banks of Fishing Creek. If there is a large flat stone near the U.S. 301 bridge over Fishing Creek, I don't see it. Maybe the rock is in the river covered up by the high water. Maybe it is actually a half mile or so from the bridge. Maybe the rock has been moved or destroyed in rebuilding the bridge. Whatever. Mark it down as just another petroglyph that got away. Another of many that have eluded this petroglyph hunter.

So now I'm walking along the bike path that parallels Western Boulevard near Pullen Park in Raleigh. Up ahead, across the road, I see the walls of Central Prison. But I keep looking over to my right—I'm looking for any well-trodden or faint path that will lead down to Rocky Branch—the little stream that runs parallel to the bike path on its south side. I've found another petroglyph reference to hunt down. This one is supposed to be on a large stone between Dorothea Dix Hospital and Central Prison on Rocky Branch, and the inscription is supposed to be: "1865 Wilson Dixson Co. C 1st no. Eng." It would have been carved by soldiers in General Sherman's army in the spring of 1865 during the several days the army was camped in and around the hills near where the Farmer's Market is found at present—on Centennial Drive just south of downtown. Union soldiers must have wandered down to Rocky Branch from their campsite for a cool drink and idled by the rocky banks. It was after the Battle of Bentonville and before the final surrender at Bennett Place, that the Union's 14th Corps, seventeen thousand men strong, were in this camp. They were there on April 17, 1865, when Sherman informed them that Lincoln had been assassinated. Raleigh residents feared they would suffer the destruction that Sherman's army had inflicted on Atlanta and Columbia but somehow Sherman and his officers restrained the men and Raleigh was spared.

It is often the case that unmarked points of interest—like Civil War petroglyphs—have well-worn dirt paths that lead off the main trail or road to the site. So at every place where I think I can discern a definite or faint footpath going down to the stream, I follow it. It's never more than a few yards. And I scan all the rock faces and look for more paths. But I see nothing even though I check all the likely suspects until I have passed the last razor wire of Central Prison. There has been extensive roadwork on Western Boulevard. And many of the high banks of Rocky Branch have collapsed into the creek. Maybe the carving has been removed or destroyed by natural processes or the DOT. Or maybe the petro is just wary and solitary. And wily. Disguises itself as a natural rock, maybe. I guess. Or hides behind lichens. Another empty handed sortie for the petrohunter.

If you drive U.S. 64 between Highlands and Cashiers and turn south on Whiteside Mountain Road (turn where you see the sign to Wildcat Cliffs Country Club) and proceed one mile to the parking area, you can hike the two mile loop trail that goes to the top and along the ridge of Whiteside Mountain. I'm there now. The summit is 4,930 feet and its great cliffs of exposed rock rise about 2,000 feet above the valley floor. A spur off the main ridge is known as the Devil's Courthouse. It is said that in the 1950s would-be developers of this area discovered a Spanish inscription at the Devil's Courthouse. The letters in the inscription, said to be two inches high and a quarter-inch deep, read: "T.T. Un Luego Santa Ala Memoria." Some have suggested that the carving was made by one of Hernando de Soto's men in 1540, as they trekked north from Florida and then west toward Tennessee. De Soto's expedition may have followed the old Indian path that ascended Whiteside Mountain from Whiteside Cove on its way westward. Hard to say what the inscription might have meant—if it is there, because, as you must know by now, I never found it. I walked the Whiteside Mountain loop, was mesmerized by the views, followed all the side trails and wandered over several exposed rock faces but never found a hint of an inscription. Interestingly, the 1939 *WPA Guide*, although it discusses Whiteside Mountain, never mentions any such Spanish inscription on the Devil's Courthouse—which it describes as "a jutting formation on the east side of Whiteside." Other more recent descriptions of Whiteside also fail to point out any 450 year old Spanish petroglyph worth visiting while hiking the summit. Hmmmm. Another one of those elusive petroglyphs.

  The last petro hunt I'll tell you about was also the one I felt most confident about. For this one I had the word of a friend that these petroglyphs existed and that he had seen them back in the 1980s. I went looking for them but found nothing. I left several petulant and increasingly surly messages for my friend and a few months later he e-mailed me that he had gone out himself and located the glyphs. They were still there, he said. His directions weren't that specific but the area to search just wasn't that big. If they were there I knew I could find them—so I enlisted the help of the petrohunter's two offspring with the promise that we would find the carvings. The three of us would stalk, discover and flush the shy petros. And a dollar to the first one to sight the prey.

  These petros were likely carved by the young son or sons of the family who ran the mill at Few's Ford in Durham County in the mid-1800s. William Few bought land on the Eno River in Durham

County in 1758. Where Few's Falls tumbles over a rocky ridge in the river, Few built a mill. Below the falls was the still shallow water of Few's Ford. William Piper took over and ran the mill in 1831 and Alexander Dickson took his turn with the mill in the 1850s and sold the land in 1863. The Piper-Dickson House now stands at Few's Ford in the Eno River State Park—at the end of Cole Mill Road, northwest of the city of Durham. On the side of the Eno River opposite the Piper-Dickson House was where the carvings are said to lie. I know the area well. We walked down past the Piper-Dickson House, forded the Eno and set to examining all the flat rock faces on the banks and in the river. We went upriver past the falls. We went downriver past the ford. We went inland looking for exposed rock faces—anywhere a kid might decide to devote some significant time to scraping his initials and the date into the living rock. And we found no such letters or numbers. Nada. We didn't even scare up a crayfish under the Eno River rocks. And I'm wondering about my good, knowledgeable, petroglyph-loving friend. He says he'll have to show me himself one of these days.

But if truth be told, this unsuccessful petroglyph hunter is not really so disappointed at all these failed hunts. Simply reading or hearing that these rock carvings may exist, that they may be out there enduring the elements and the years, brings a smile to my face. And wandering up mountains, down streams and through the woods, looking for a likely place that someone might choose to sit a spell and laboriously carve out a message for the ages, is wonderful fun. Maybe not everybody's idea of fun, but it is for me. And I'm also conflicted. I'm not sure I want there to be well-trodden paths that mark the way to these rock carvings. The spray paint and other vandalism might not be far behind. Maybe these petroglyphs should remain unpublicized, rarely seen and elusive. As a matter of fact, maybe I actually did successfully stalk and stumble upon each—or some—of these petros. Maybe I've got the photographs to prove it. And maybe not telling you, and leaving you guessing whether they are there or not, and if so, where, is what's best for both the petros and my fellow petrohunters. Maybe so. But if you take up the hunt and discover one of these petroglyphs, why don't you send me an e-mail and let me know. Some general directions would be nice.

Tom Fowler

Paint Rock

Tarleton's Tea Table

In 1780 the British army confidently invaded the American South. First Georgia and then South Carolina fell to the British forces under Lord Charles Cornwallis. By the fall of 1780, Cornwallis was ready for the next step in his subjugation of the former colonies—the invasion of North Carolina. The American force that stood in the way was the rag-tag American Southern Army led by General Nathanael Greene who assumed command on December 3, 1780. After some hesitation and false starts (maybe caused by troubling defeats at the Battles of King's Mountain and Cowpens), Cornwallis' army crossed into North Carolina in the third week of January, 1781.

General Greene described his army of two thousand as a "shadow" of an army. He said his soldiers were "wretched beyond description" because of their lack of provisions, clothing and other necessities. Any pitched battle with Cornwallis' experienced, disciplined and well-supplied troops seemed out of the question. So General Greene retreated to the northwest from his headquarters in Charlotte. And Cornwallis gave chase across the Piedmont of North Carolina and all the way into Virginia. Greene's retreat turned out to be an inspired strategy. Despite some close calls, Cornwallis never caught up with Greene's forces. But his unsuccessful pursuit of Greene across North Carolina's midsection seriously weakened the British. Once in Virginia, Greene's army was reinforced and the Americans returned to North Carolina to fight Cornwallis at the Battle of Guilford Courthouse on March 15, 1781. Although technically the victor at Guilford Courthouse, Cornwallis' army was never the same afterward. Demoralized and weakened, the British retreated to Wilmington and then they marched north—north to the final British defeat at Yorktown in October 1781.

Cornwallis' pursuit of the American Southern Army in the winter of 1781 was led by the swift green-coated British cavalry, under the command of the infamous Lieutenant Colonel Banastre Tarleton. Tarleton was a cocky, red-haired, short tempered twenty-six year old Oxford graduate, who displayed both military genius and an appalling disregard for basic humanity. Though sophisticated and charming, when he chose to be, Tarleton also allowed his troops to slaughter surrendering American prisoners. This earned him the nickname of "Bloody Ban," and caused American reprisals on British prisoners in the name of providing "Tarleton's Quarter" (which meant no quarter given).

Tarleton was also notorious for his rude treatment of the local civilian population, including his commandeering of needed supplies and shelter, and his destruction of what he could not carry off. As Greene's army retreated from Charlotte, to Salisbury, Mocksville, Old Salem and Hillsborough, the terrified civilians left behind would cry "Tarleton is coming" and hide their valuables. On June 4, 1781, Tarleton's cavalry raiders came within ten minutes of capturing Virginia Governor Thomas Jefferson in Charlottesville, Virginia.

Early in its pursuit of Greene, Cornwallis' army camped for three days near present day Lincolnton on the site of an earlier battle at Ramsour's Mill. Lieutenant Colonel Tarleton was served his meals—and possibly his tea—on a large, relatively flat, boulder at the campsite. This rock, known as "Tarleton's Tea Table," now sits on the grounds of the Lincoln County Courthouse in Lincolnton. It's on the north side of the courthouse grounds at the edge of the lawn, and it is marked with a D.A.R. plaque. It's as good a place to have your tea as it was in 1781, and nowadays you won't have to make room at the rock for a tea enthusiast called Bloody Ban.

General Cornwallis, also a cultured Englishman, looked for natural tea tables too—he preferred not to neglect this afternoon tradition. After Cornwallis' army decamped from Lincolnton, it marched eastward through Lincoln County towards the Catawba River and Cowan's Ford (now the site of a huge hydroelectric dam near N.C. Highway 73). The British army crossed Cowan's Ford on February 1, 1781, under heavy fire from the American army in a rearguard action. British losses were high—thirty-one killed and thirty-five wounded. General Cornwallis' horse was shot out from under him during the crossing. The Americans gained valuable time by impeding Cornwallis' crossing. But the day before the crossing at Cowan's Ford, Lord Cornwallis stopped at another large, flat rock in Lincoln County for his afternoon tea. This rock, known of course as "Cornwallis' Tea Table," is located about a half mile north of N.C. Highway 73, several miles west of the intersection of Highways 73 and 16. It endures unmarked and covered with lichens, and has probably not hosted a tea in many a year.

Tarleton Tea Table Rock

The View from the Firetower on Flat Top Mountain

Since the late 1970s, at regular three or four year intervals, I find myself standing at the top of the firetower on Flat Top Mountain taking stock of what I've done and what I want to do in this life. Sometimes it's winter and I've cross-country skied to the top. Sometimes it's summer and I've jogged the trail to the tower—and sometimes I've just walked it. I always stand at the top, hands on the rail, and slowly scan the 360 degree view, from Grandfather Mountain to the west, and the town of Blowing Rock to the south. I stay there long enough—I'm usually there all by myself—until I compose, and say out loud, what I call a "Flat Top Firetower Statement." The statement is always about what I'll get done before I next climb the carriage trail up to Flat Top Mountain and stand atop the firetower. And maybe I don't really say it out loud—but it always feels like I do.

The trail up Flat Top Mountain begins at the parking lot for the grand old mansion that houses the Southern Highlands Handicraft Guild on the Blue Ridge Parkway between mile posts 292 and 295. It's also just up Highway 221 from Blowing Rock. Moses Cone, a giant of the textile industry in the late 1800s, built the house which he called "Flat Top Manor." Cone and his wife bought up much of the surrounding land and built 25 miles of interconnecting carriage trails that meander over the vast estate, including the one that goes to the summit of Flat Top, at 4,558 feet above sea level. Cone died in 1908. In 1950, Cone's widow, Bertha, deeded the estate to the National Park Service—it became Moses H. Cone Memorial Park, 3,517 acres of field, forest and mountain. The carriage trails that run throughout the property were built to have only gentle grades—making for consistently mild to moderate climbs, easy on the hiker or runner, and ideal for the cross-country skier.

Walk down the steps from the Manor House parking lot to the gravel road that leads toward the Carriage House. Turn left on the road and cross under the Blue Ridge Parkway. At the first intersection, bear to the right—this is the Flat Top Carriage Trail (to the left is the Rich Mountain Carriage Trail). You will ascend along the old road with the forest on your right and a pretty field on your left. A switchback to the right will take you into the forest and after about three quarters of a mile you will enter a large open field. At about the one mile point on your left you will see the Cone family cemetery, where Moses and Bertha Cone are buried. It is a lovely spot and an appropriate place to think of the Cones and to thank them for preserving the land for our use.

The carriage trail continues across the broad field which slowly rises up the flanks of Flat Top toward its forested upper slopes. You enter the forest at about 1.5 miles and begin a series of twists and switchbacks up through the forest. It can be muddy in places with pools of water if it has rained recently. But the ascent is gradual and the forest lovely. The summit and the firetower are reached at about the three mile mark. And after you have climbed the tower, surveyed the high country in all directions and considered and crafted your own statement—if such is your wont—then it is three miles back down to the parking lot.

When I climb up the steps of the firetower and finally stand at the top, I never can recall much of the details of the last Flat Top Firetower Statement. But even if I didn't feel so great about life at the trailhead, by the time I'm on top of the firetower I'm usually feeling pretty good. And pretty optimistic. I gaze over at Grandfather Mountain and wait until the latest statement forms in my mind. Then I either say it, or think I say it, take a last 360 view and head down the tower, down the carriage trail, past the Cones' resting place, and back into the life of a flatlander. Until next time.

For more information on Moses Cone and his time on Flat Top, see Philip T. Noblitt's *A Mansion in the Mountains: The Story of Moses and Bertha Cone and their Blowing Rock Manor* (Parkway Publishers, 1996)

Snowy field atop Flat Top Mountain

Peter Ney

In the fall of 1819, a veteran soldier of Napoleon's army, who had left France for a life in America, was watching a parade in Georgetown, South Carolina. As a group of horsemen pranced by, the soldier recognized one of Napoleon's highest ranking officers—a man Napoleon had once described as "the bravest of the brave." It was Marshal Michel Ney, known to his men as "Red Peter," because of his reddish-blonde hair and florid complexion. Marshal Ney had commanded the Third Corps of Napoleon's Grande Armèe, and had fought in the Swiss Campaign, the Russian Campaign and the Battle of Waterloo. Ney had been instrumental in Napoleon's brief return to power in 1815 after exile to Elba. So it was somewhat surprising to see Marshal Ney in South Carolina. But what made it particularly strange was that Marshal Ney had been dead for four years. He had been executed by a firing squad in Paris on December 7, 1815.

Marshal Ney had been executed for treason because of his help in returning Napoleon to power. Reports at the time stated that the firing squad consisted of twelve riflemen, that Marshal Ney saluted them and gave the order to fire himself, that the body lay on the ground for fifteen minutes as the law required, and that the body was then taken away and buried in the Père la Chaise Cemetery in Paris. Sounds pretty straightforward—but it was also reported that the twelve riflemen of the firing squad were all veterans who had served under Marshal Ney during the Napoleonic Wars and that the time and place of the execution were changed at the last minute so that the large crowd that had assembled to witness the event were at the wrong location.

At the actual execution site in Luxembourg Gardens, there were few spectators and those present were not even aware of who the person being executed was. A few days later a body said to be Ney's was buried at Père la Chaise but it is interesting that Ney's widow did not attend the burial and the family never placed any memorial on this grave.

Was Ney's execution faked? Possibly. Like many of the high-ranking officers in the armies of the various European countries, Ney was a member of the ancient fraternity of Masons. The Duke of

Wellington was a Mason. Masons often protected and aided each other even when they fought on different sides of a war. And both Napoleon and Ney retained broad support among their former soldiers and the general French populace, even after Waterloo and the restoration of the monarchy. It is also interesting that when Ney was sentenced to death by France's Chamber of Peers, one of the sentencing options considered was exile to America—and it is reported that of those who voted for death some nevertheless were not comfortable with the death sentence and considered exile to be acceptable. It is possible, then, that after a faked execution Marshal Ney was spirited away to Bordeaux where he boarded a ship headed for Charleston, South Carolina, and relative obscurity.

There might not be much more to this story except for a memorial found in the cemetery of Third Creek Presbyterian Church, a church built in 1835, near the town of Cleveland in Rowan County. The headstone on this grave is inscribed: "In Memory of Peter Stewart Ney a native of France and soldier of the French Revolution under Napoleon Bonaparte who departed this life Nov. 15, 1846, aged 77 years." Many who knew Peter Stewart Ney believed him to be Marshal Ney.

Peter Stewart Ney impressed most everybody who met him. He was routinely described as the "greatest man" they knew. From 1820 until his death in 1846, Ney taught school in various small communities in the Piedmont areas of North and South Carolina, and Virginia. But he was much more than a schoolteacher. He was a scholar, speaking several languages fluently (including French). He was particularly adept at mathematics. At the request of the Board of Trustees, Ney designed the official seal (still in use today) of Davidson College. He wrote poetry and contributed to several local newspapers. He was also acknowledged as a master swordsman and horseman. Even in his 60s and 70s he was described as agile and athletic, and always comporting himself with a distinctive military bearing. And Ney had significant scars of combat all over his body—including a deep sabre scar on the left side of his head, sword wounds on an arm and a thigh, and a musket ball in his calf.

The physical similarities between Peter Stewart Ney and Marshal Ney (had he lived into his 50s, 60s and 70s) were great.

They would have been about the same age, same size, coloring and build. Marshal Ney was an expert fencer and cavalryman. And several former soldiers who had known Marshal Ney in Europe upon seeing Peter Ney in America proclaimed him to be the marshal.

Perhaps more convincing, and certainly more moving, are the reports of conversations and experiences Peter Ney had with various acquaintances in and around Rowan County, North Carolina. Although usually reticent to discuss his past (he often said only that he was a French refugee and had fled France for political reasons), to some he revealed details of Napoleon's campaigns and commented on many of the famous personalities involved in the Napoleonic Wars. Peter Ney also wrote comments in the margins of books about Napoleonic history, noting inaccuracies and making corrections. Ney was absolutely devoted to Napoleon and was devastated upon learning of Napoleon's death in 1821 and again upon Napoleon's son's death in 1832. Peter Ney had confided to some that he expected to return to France when the French people returned a Napoleon to rule over France once more. The deaths convinced Ney he would never be allowed to return to France—where Marshal Ney's widow and children still lived.

As he lay dying in November of 1846, Peter Stewart Ney was asked to say "who you are before you die." He reportedly answered "I am Marshal Ney of France." He died several hours later. Ney's very last words, uttered while not fully conscious, were: "Bessieres is dead and the Old Guard is defeated. Now let me die." Marshal Bessieres, the beloved commander of the Old Guard, died in battle next to Marshal Ney in 1813, and despite many victories under Bessieres and Ney, the Old Guard had been defeated at Waterloo.

Much research has been done to investigate the connection between Marshal Ney and Peter Stewart Ney. Much of it (including handwriting analysis) is supportive of this connection. In 1887 the body of Peter Stewart Ney was exhumed and physicians examined the remains for evidence that they expected to find in the remains of Marshal Ney—but they found nothing conclusive. There may be no more evidence to find. So the mystery may remain forever unresolved. Maybe it is best left as one acquaintance of Peter Stewart Ney said years after Ney's death: "Well, if he was not Marshal Ney, he ought to have been."

Directions to Ney's Grave: The Third Creek Presbyterian Church cemetery is northwest of the town of Cleveland. Follow Third Creek Church Road out of town for just over a mile and a half, and look for a left turn on S.R. 1973. The church is on the right in about a quarter mile.

Peter Ney's grave site

The Night of the Shooting Stars

It was early in the morning when my radio clicked on. I listened for awhile in the dark. I was about to get up and head for the bathroom when the announcer said they had awarded the 2002 Nobel Prize for Physics to three guys. I paused and lay back down to listen. I'm no physicist, and I've never even taken a single course in physics, so it may surprise you to know that I've actually met some Nobel Prize winners in physics—met them before they ever became Nobel laureates. I wondered if it could happen again. The radio said that the winners were Masatoshi Koshiba, Riccardo Giacconi, and a Dr. Raymond Davis, professor emeritus at the University of Pennsylvania. Dr. Davis won for his research on neutrinos— little subatomic particles produced in the nuclear fusion reactions occurring in the core of the sun that shoot out across the universe passing directly through everything they encounter (neutrinos rarely interact with matter). In the 1970s and 80s, Dr. Davis marked the passage of these solar neutrinos using chlorine detectors located deep underground in a South Dakota gold mine. Yessir, Dr. Raymond Davis. We knew him as Ray.

Back in the 1960s, my father, a physics professor, worked for a time at the Brookhaven National Lab on Long Island, New York— the same place Ray Davis worked. Our families got to know each other and we used to go sailing with Ray Davis on his 25 foot sailboat. I remember one day in particular when we were out on his boat, tacking across Long Island Sound. A sunny day and a stiff nor'easter. Ray and I were back in the cockpit just shooting the breeze and I recall saying to him, "Ray, it might be interesting to take a look at this neutrino business." He just nodded, sipped his beer and stared out at the water. But I guess he was listening.

That's how I remember it, anyway. Yeah, it's true that I was just a kid back in the 60s. But that doesn't mean I wasn't on the ball. I paid attention when my father talked about particle physics, detectors, scanners, cloud chambers, bubble chambers and accelerators. And yeah, maybe I really thought that neutrinos was a breakfast cereal. It was a long time ago. And also, I would have liked to have known Norma Jean—but I was just a kid, as I told you. But as kids grow up, they listen, they take mental notes and then they start asking the tough questions. Particularly when their father is a high energy physicist who took his family with him when he did experiments on the big atom smashers, like CERN in Switzerland, Fermilab in Illinois, and SLAC which straddles the San Andreas Fault in Palo Alto, California.

So I had, at least some understanding that atoms were not the fundamental building blocks of nature—that the protons, neutrons and electrons that make up an atom are actually made up of even smaller little things, and that physicists learn about these even smaller little things by smashing atoms together and watching what the resulting little pieces do after the impact. I found this mildly interesting. And I remember my professor pop talking about Gell-Mann and Feynman, and a host of other theoretical physicists. Some would stop by our house to visit pop, and some would stay the night. A few would later win their Nobel. So after hearing about quarks for most of my life, and occasionally asking my father for explanations, as a teenager, I asked my dad, "So, Pop, 'zup with them quarks?"

Theoretical physicists are often well-rounded types, my father opined, and when, in the 1950s, Caltech's Murray Gell-Mann first proposed the existence of the tiny little particles that really were the fundamental building blocks of matter, he chose to call them "quarks." Why? Well, the word came from James Joyce's phrase "Three quarks for Muster Mark," in *Ulysses*, says my pop. Gell-mann said there were three kinds of quarks (up, down and strange) and Joyce postulated three quarks, so there you go, said my dad. Now that was sort of intriguing to a well-rounded political science major, as I was in the mid 1970s. I resolved to check into it at some point. But I was in no rush to figure out this subatomic particle zoo that was the business of high energy physics. I'd let it play out for awhile on its own. So in the 1980s and 1990s I paid scant attention to developments in particle physics. Sure, I thought about black holes and read Stephen Hawking's book, *A Brief History of Time*, Riordan's *The Hunting of the Quark*, and finally *Driving Mr. Albert* (about a cross-country trip with Einstein's brain). But then finally, in the mid 90s, I was ready. Ready to match, or at least follow, the intellectual path of the man who first perceived the ghostlike existence of the quark, Dr. Gell-Mann. I bought a copy of *Ulysses* and started reading.

It wasn't easy and it wasn't always pretty, but I finally plowed through all of Joyce's *Ulysses*. And it wasn't there. No quarks at all for Muster Mark. And no Muster Mark either. Much to my disappointment and then to my outrage, I never found that phrase about Muster Mark and his quarks. So I confronted my father yet again about inconsistencies in his physics. "There ain't no quarks in there," I screamed, waving the book in his face. Roused from pondering string theory in the far reaches of the universe, my father, with an effort, raised his head to focus on me and my issue. "No, no, son," he soothed, "the three quarks for Muster Mark was in *Finnegan's Wake*, not *Ulysses*." Well, dad, thanks for clearing that up. *Ulysses* ... *Finnegan's Wake* ... it doesn't really matter, eh, son? That's just Gell-Mann! Grrr. Stupid non-literary physicists—conflating *Ulysses* and *Finnegan's Wake*. Guess some time in the next thirty years I'll get around to plowing through *Finnegan's Wake* as well. Those quarks better be there, though. Or maybe, as I do on my father's claim that he understands Einstein's theory of relativity, I'll just have to trust him on this one.

So for a time I refused to think about particle physics, quarks, neutrinos, wormholes and the like. But then I heard about the meteor shower.

Hope Valley Road was pretty well deserted—as might be expected at 4:20 a.m. early this Sunday morning. My wife and I had our travel coffee mugs, and my son had his hot chocolate, so we were relatively content as we drove south from Durham headed for the darkness of Jordan Lake. A few days before, I had told them I was going to get up at 4:00 a.m. Sunday to go see the Leonid Meteor Shower. I smiled and asked if they wanted to come. Much to my surprise they both said yes. So the alarm was set, the sleep-deprived were awakened, we got our chosen beverages and we started up the car and drove away through the sleeping city.

The fall had been remarkably mild and clear. Bad for our water supply but the consecutive weeks of high pressure kept the clouds away and ensured that Sunday morning would be calm and cloudless—ideal for star gazing and meteor watching. The newspapers and television had told us the Leonid Shower should be a big one— multiple shooting stars every minute. And if it was clear and dark, the night sky should be spectacular. That is why we were headed out to Jordan Lake—to get away from the city lights and to look out over the dark waters. The official viewing spot was to be Ebenezer Church Recreation Area. We figured we'd have the place mostly to ourselves.

Crossing over Highway 54 we saw some red taillights up ahead. Maybe some other folk were up for some meteor viewing after all. As we sped down Highway 751, headlights appeared behind us. Soon we formed a little caravan, backed up behind a slow moving van. We didn't expect traffic problems on this drive but it was encouraging to know that a few others were making the effort to see the shower.

At the first bridge over Jordan Lake both shoulders of the road were filled with parked cars. Meteor watchers? It had to be. And yes, we could see people standing next to the guard rails. We looked up through the windshield and saw our first shooting star of the morning. Every time we drove clear of the trees we would see the dark sky and another meteor. There were more cars parked on the sides of the road at every bridge.

As we approached Highway 64, the traffic backed up. We stopped, then gradually crept forward. What was going on? As we got up to 64, we looked to our left, toward Raleigh, and saw a constant stream of headlights headed west, toward Jordan Lake. We turned onto 64 and joined the line of cars. A few miles down the road, the left turn to Ebenezer Church Recreation Area was backed up for over a quarter of a mile. We'd never get in, I thought. Thousands of people had left their warm beds to come out and see the meteors and to clog the roads around Jordan Lake. I was impressed and amazed. I had no idea so many others would be drawn to the shower. But I was also concerned that the crowds had filled the viewing spots. We kept driving west on 64.

All the recreation areas we passed were closed with their gates locked. We came up to the causeway that crosses the middle of Jordan Lake. Cars were parked on both shoulders and in the median. At the first empty stretch of grass on the right, I slid the car in and parked. We got out and leaned against the car with our backs toward the traffic on 64. The van parked in front of us and the SUV parked behind us blocking out the passing headlights. We looked out across the lake and the stars began falling.

They came in ones, twos and threes, through all parts of the sky. And they came constantly. Particularly bright ones would leave a smoky white trail that would linger across the sky before evaporating. Our small party, and the groups gathered on either side of us, would gasp and remark when a big meteor or a double or triple shot across the heavens—just like at a Fourth of July fireworks display. But in between, we were quiet, awed by what we were seeing. Distant meteor trails flashed regularly over the far horizon but the brightest ones were those we saw looking straight up into the night. They all seemed to be shooting toward the west. Gradually, we stopped exclaiming at particularly dramatic meteors, and we simply watched in total silence. Far up in the sky, I could identify a few constellations and their twinkling stars, motionless behind the falling stars. It was impossible not to think of these stars as simply enduring in place, and waiting for their turn to blaze across the sky and disappear into the void.

The meteors fell, time passed and a soft wind blew across the lake. When the SUV parked behind us turned on its headlights and drove away, we snapped out of our reverie and realized that we were a bit chilled and we had cricks in our necks. With the big SUV no longer blocking the headlights from the traffic on 64, we were distracted from the sky and someone mentioned breakfast. There was still no glow in the East and the stars were still falling. But the expected warmth of the car was compelling, we got in, turned up the heater, and decided to join the exodus away from Jordan Lake.

We drove back the way we had come earlier that morning, past many cars still parked on the shoulders and a few late arriving meteor watchers. We were warming up and the sky was light when we pulled into the parking lot of Biscuitville. But Biscuitville was not yet open, so we drove back to our still sleeping neighborhood to complete our reentry into our earthbound existence.

So I pull into our driveway and shut off the car. We gather up our coffee mugs, lock the car and walk to our porch. At the door to our house, just as I slip the key in the lock, three neutrinos slam into the bed of azaleas in our front yard. They burrow on through the earth's crust, mantle and core, and shoot out the other side, headed for deep space. In a great hurry, as always. Then, from somewhere in my memory, an image flashes in my mind of my father and Ray Davis, much younger, in bathing suits and canvas shoes, sitting together in the old sailboat, Ray with his hand upon the tiller. They are smiling, laughing about something, and the boat heels over in the wind. Then they turn to look my way. And my father winks at me. Against all probability (as I have never been the winking type), I actually wink back. "Three quarks for both of you musters," I think to myself. My wife, watching me, smiles and ushers me inside.

The Forks of the Yadkin

It's Dan'l Boone and Kentucky, right? Boone—that Indian-fighting frontiersman, who cut out the Wilderness Road through the Cumberland Gap, and who led parties of settlers along this path to populate the great blue grass state. Boone founded the fortress town of Boonesborough on the Kentucky River and defended it against Indian attack throughout the 1770s and 80s. Boone carved his name on trees (*note*: a famous carving discovered on a beech tree on the banks of the Watauga River read: *"D. Boon CillED A. Bar on tree in the YEAR 1760"*) and was a legendary bear hunter. Once he was captured by the Shawnee, taken far north to their Ohio town of Chillicothe. Boone lived with the Shawnee for a year and was adopted as a son by the local chief. He later escaped and traveled hundreds of miles by foot to Boonesborough to warn of a planned Shawnee attack. Boone later served as a legislator and was the subject of a popular book about frontier life in America—John Filson's *The Adventures of Col. Daniel Boone* published in 1784. In all his ramblings through trackless miles of wilderness, Boone claimed he'd never been lost a day in his life—though he allowed as how once he had been a tad "bewildered for three days." And, of course, as many of us recall, Dan'l Boone had his own television show in the 1960s. But this famous man will forever be associated with Kentucky and its settlement by those of European descent in the years before and during the Revolutionary War. It may not generally be known, however, that from age sixteen until his early forties, Daniel Boone was a North Carolinian.

Boone was born in Pennsylvania in the fall of 1734, the sixth of eleven children. In 1750, Daniel's parents, Squire and Sarah Boone, moved the family down the Great Wagon Road to the North Carolina Piedmont, to an area where two prongs of the Yadkin River come together—an area known as the Forks of the Yadkin. Local tradition tells us that at first the Boones stayed in or near a cave on the eastern bank of the Yadkin in Davidson County. Boone's Cave, once a state park and now leased to the county for use as a park, is found fourteen miles west of Lexington, off of Highway 150. Look for signs for Boone's Cave Road. A few miles downstream from Boone's Cave is where the South Fork of the Yadkin splits off toward the west and the Brushy Mountains. The other branch of the Yadkin heads north (upstream) past Winston-Salem until it turns west and forms the boundary between Surry and Yadkin Counties. This branch of the Yadkin flows through Elkin and Wilkesboro. The Yadkin's source is found in Blowing Rock. Daniel Boone would come to know these rivers, and the land in between them, very well.

Squire Boone purchased hundreds of acres in the Forks of the Yadkin, near the present-day town of Mocksville in Davie County. West of Mocksville U.S. 64 crosses over Bear Creek. A state historical marker just before the bridge proclaims: "Boone Tract: In 1753 Lord Granville granted 640 acres on Bear Creek to Squire Boone who sold it in 1759 to his son Daniel. This was part of the original Boone tract." During this time, the elder Boone children were marrying and starting their own families. It is said that although Daniel's help, as the oldest child remaining at home, was needed on his father's farm to clear and plow land, and to tend the crops and livestock, Daniel "never took any delight in farming." Daniel much preferred the life of a hunter. An excellent shot, Daniel would disappear into the woods for days, shooting deer and bear, and then pack his deerskins for trade in the tiny commercial hub of the region—the crossroads town of Salisbury in Rowan County. But Daniel stuck around home long enough to court one of the daughters of a nearby family.

Daniel Boone, age twenty-one, and Rebecca Bryan, age seventeen, were married on August 14, 1756. They moved into a cabin on Sugartree Creek about two miles east of present day Farmington in Davie County. And by the time Rebecca turned twenty, she and Daniel had four children. More would follow. Daniel and Rebecca lived at the cabin on Sugartree Creek for about ten years. Daniel continued his wandering ways, going on hunting trips for weeks and months at a time. In 1760 he crossed the Blue Ridge for the first time, in 1761 he fought with General Rutherford in the campaign against the Cherokee, and in 1765 he explored Florida. His wandering may have been encouraged by financial problems and the death of his father. Court records in Rowan County show a judgment against Daniel Boone in 1764 for the substantial sum of fifty pounds. Daniel's father, Squire Boone, died in 1765. Squire and Sarah Boone are buried in the Joppa Cemetery northwest of Mocksville on U.S. 601 (park in the shopping center parking lot next to the cemetery).

Whatever the reason, in 1766 Daniel and Rebecca moved their family many miles up the North Yadkin River to Holman's Ford near where Highway 421 crosses the Yadkin south of Wilkesboro. Soon afterward they moved again, this time building a cabin further upriver on the north side of the Yadkin on a hill overlooking the river and opposite the mouth of Beaver Creek, near the present-day town of Ferguson, in Wilkes County. The remains of this cabin were visible well into the twentieth century. Rebecca raised the children and ran the home place while Daniel explored. He reached Kentucky for the first time in 1767 and for a second time in 1769. It was on this latter trip that Boone was captured by the Shawnee and taken to Ohio. It would be two years before he returned to North Carolina, Rebecca and the cabin on the Yadkin. But Boone persisted in returning to Kentucky, taking settlers and occasionally his family with him. Rebecca left Daniel and Kentucky to return her family to the Forks of the Yadkin on at least one occasion, but finally left North Carolina for good in 1778—to reunite the family in Kentucky.

There are many other sites in North Carolina associated with Daniel Boone—various gaps he passed through, springs he used, and, of course, the town of Boone where Boone kept a hunting cabin in the 1760s. There is also much literature trying to retrace the various paths or trails that Daniel Boone followed. Old U.S. 421 west of Boone (leading north to Zionville and Trade, Tennessee) is said to follow Boone's Trail and has several historical markers commemorating Boone's route. One of the most intriguing sites to visit is the cemetery near Obids off Highway 163, fifteen miles or so northwest of Wilkesboro (follow Highway 16). In the Calloway Cemetery is the gravestone for Captain Thomas Calloway. Surmounting this monument is a stone shaft that Daniel Boone used as a campsite marker—and inscribed on this marker are the initials "T.C." It is said that Boone carved those initials himself and placed this stone on the grave of his friend Thomas Calloway.

Boone lived a long, productive and exciting life as he followed the frontier as it advanced ever westward, always being harried from the rear (or east) by new homesteads and advancing civilization.

Boone's adventures in Kentucky and Ohio are well worth discovering—but we can also walk in the footsteps of the young Daniel Boone in and around the creeks and woods of the Forks of the Yadkin.

Postscript: Even Kentucky finally became too settled and crowded for Daniel Boone. Boone continued his life-long westward migration by moving his family to Missouri in 1799—he died there on September 26, 1820. Although initially buried there his remains (and Rebecca's) were disinterred and reburied in Frankfort Cemetery, Frankfort, Kentucky in 1845 (although there is some debate as to whether the correct body was exhumed). For further information about D. Boone, his life and times, see:
- John Mack Faragher, *Daniel Boone: The Life and Legend of an American Pioneer* (Henry Holt and Co., 1992)
- George H. Maurice, *Daniel Boone in North Carolina* (Murray Printing Co., 1955). This book has maps and photographs of the various Boone home sites in North Carolina.

Squire Boone's grave

After the Olympics: Dr. Godiva Ponders The Future of Running in America

It was a hot October night in North Carolina. I'd just gotten back from a three mile evening run. It should have been five but the extra loop just didn't seem worth it tonight. The Olympics were over and nights were getting back to normal. The kids and my wife were already in bed. I was sitting, dripping in the kitchen when the phone rang. It was Dr. Ernest Godiva, the sage old runner and self-appointed spokesman for our local running club—and my long-time friend. He appeared to be having some sort of fit—"very, very concerned" as he put it with that faux Viennese accent of his. Come on over, I told him, knowing he was probably already well on his way. So I ambled over to the old Kelvinator, mined two Coronas from a narrow vein I had prospected earlier in the week, found a reasonably good looking half o' lime behind the sugar bowl on the countertop, sliced it up, balanced the frog-colored wedges on each bottle top, and waited for the doc. In no time at all, the door was flung open and in marched the agitated Dr. Godiva. I handed him a Corona, and prepared for one of the old fellow's inconclusive stories. The doctor sighed heavily, took a long draught from the bottle, and leaned forward in his chair. I leaned back in mine and as I did so heard a roll of thunder far off in the distance. In a low voice, and with great intensity, the doctor began his disturbing tale—its gist, as follows:

> *Just like I warned you, it has happened. Maybe it is not quite the end but the last act is surely well under way. The U.S. men's soccer team made it all the way to the medal round at the Olympics in Sydney for the first time ever— and the U.S. women's soccer team, long dominant in the world, was barely edged out for the Olympic gold. Meanwhile American distance runners were absent from most of the Olympic finals and were invisible during the marathon coverage. You say, "So?" You foolish, foolish person!! While America's youth flocks to soccer, America's distance runners are aging, slowing and disappearing from the world scene. Once world class, American distance runners are now second class, at best. Our young folk are*

Tom Fowler

now playing soccer in huge numbers and are competing with the world.

With no new influx of young runners, the average age of road racers is increasing, as is the average age of running club members. In local races the over 35 and over 40 age groups are always the largest. The overriding importance of age group awards, the increased popularity of 5K races at the expense of longer distances, baby jogger use and complications, the increased popularity of alternative events: age-handicap races, hashing, ultras, orienteering, walks, etc., are some of the consequences. In-line skates, cross-country skiing, mountain bikes, bungee-jumping, how can running compete? It is hopeless, hopeless, hopeless. Our running club will soon become simply a retirement activity and will lobby against social security cuts and estate taxes. ... 5K Run for Medicare ..... cost of prescription drugs ...... AARP .... Geritol ads in the newsletter ....

Thunder crashed nearby and I awoke with a start. Had I been dreaming? No. The Coronas were upright but empty on the floor, the frog-green wedges mashed and safely inside the bottles. Ernest was gone and the house was quiet and dark. The rain began drumming on the roof. My knees ached a little as I stood up. Maybe I needed to pick up my mileage a bit, I thought to myself. Get back in shape. Run a 10K. Maybe a marathon? Walking into the living room I noticed my four year old son's soccer ball on the sofa. I picked it up, tossed it high in the air and caught it. I stood there a moment lost in thought. Then I walked to the closet and placed the soccer ball high up on the closet's top shelf and hid it behind the box of Legos.

The future of running in the United States

Judaculla Rock

The intrepid petroglyph hunter, finding himself in Mesa Verde National Park, is inexorably drawn to the trail that is known as the Petroglyph Point Trail. But it's a long trail, with no signs marking said petroglyphs, no obvious points, and lots of smooth rock faces that would seem a perfect home for those rock carvings known as petroglyphs. But we can discern no man made rock art on the canyon wall beside the trail. And the hunter is hurting. He's been nursing a bum knee for several months now. So with his two boys, aged 9 and 11, disappearing down the trail, the hunter shouts that there will be a dollar for the first boy to spy a petroglyph. This reinvigorates their fragile interest in rock art and they press ahead. We hike on another half mile or so, turn a corner and there they are—unmistakable, dramatic and mysterious. There are carvings of mountain goats, human-like figures, geometric designs and much more. They have persisted on this rocky face for several hundreds of years and maybe much longer. Well worth the dollar, in my opinion.

Petroglyphs are inscriptions or carvings on naturally occurring rock faces. Native Americans have left thousands of these carvings across the United States. Some may be a thousand years old while others may be only a few hundred years old or less. Rock art also includes pictographs—which are paintings on naturally occurring rock. Many pictographs survive in the arid Southwest but the rock paintings are much more fragile than the rock carvings. The meaning and purpose of this rock art is much debated by anthropologist and amateur rock art hunter alike. Petroglyphs sometimes clearly portray animals or hunting scenes, but many carvings are of hand prints, foot prints, and bizarre human like figures that are not easily interpreted. And many petroglyphs are of a great variety of geometric symbols or figures. Are petroglyphs writing or record keeping? Do they have a religious or ceremonial connection—are they a shamanistic vision? Or are they simply art? Or maybe rock art is all this and more. We may never know.

The petroglyphs that most of us have seen are those found in the American Southwest. Relatively few have been found in the eastern states. But petroglyphs do exist throughout the United States.

There may be more rock art in the Southwest than in the East but it is also true that there is simply less exposed rock in the East. Particularly in the Southern states, the vegetation and high humidity may hide and disguise the rock art that does exist, and both probably contribute to the natural deterioration of both pictograph and petroglyph. Additionally, the greater population concentrated in the East and the resulting construction and development has probably destroyed the natural habitat of eastern rock art or at least limited the petroglyph hunter's access to or knowledge of the site. It does seem likely, however, that there are numerous lichen-covered petroglyphs in the eastern states that are waiting to be rediscovered.

South Carolina may prove this point. In the late 1990s, South Carolina's Institute of Archaeology and Anthropology initiated a "Rock Art Survey." This survey used volunteers to walk over certain tracts of land looking for petroglyphs and it also used media outlets to ask the general public to report any possible rock art sites. In its early stages this survey resulted in the recording of approximately 150 rock art sites in South Carolina.

Although the professional anthropologists may know differently, there are few publicly acknowledged petroglyphs to be found in North Carolina. But there is one very famous one. It's located in Jackson County near Cullowhee and it's known as Judaculla Rock.

To visit this rock, get yourself to Cullowhee (where a Native American effigy mound located on the campus of Western Carolina was excavated in 1898 and leveled by campus construction in 1956) and then travel south on N.C. 107 for about three miles. Turn left (east) on Caney Fork Road (S.R. 1737)—at this intersection you'll also see a silver state historical marker for Judaculla Rock. Follow this road, paralleling Caney Fork Creek, for about two and a half miles and you'll see a sign for Judaculla Rock on the left—turn left on this gravel road. The road winds through a field gradually turning to the right—after about a half mile you will look to your right down a short slope and see the rock and viewing platform. Park and walk down to the rock.

Judaculla is a big rock and it is absolutely covered with rock carvings. Many of the carvings are cuplike depressions in the rock, known as cupules—possibly used for ritual purposes. There are also many deep grooves in the rock. And then there are the designs and figures, some of which look like footprints, humans, humanoids, birds and some that just look fascinating and mysterious. Some researchers claim to have discerned on Judaculla Rock the famous Kokopelli figure depicted in many petroglyphs found in the desert Southwest. The Kokopelli figure is said to be a hunch-backed, dancing flute player, and, indeed, there is a Judaculla carving that could be viewed as such.

Researchers in the late 1800s concluded that those who created the petroglyphs on Judaculla Rock were from a culture that pre-dated the Cherokee. James Mooney, one of these researchers, nevertheless gathered this story from the Cherokee: Judaculla (Tsul-ka-lu) was a great slant-eyed giant who lived on the top of a mountain above the Tuckaseegee River. One time Judaculla jumped from his mountain top and landed on the rock, leaving the imprints that we see today. One of the petroglyphs on the rock does somewhat resemble a seven-toed foot and the giant may have had seven toes.

The sheer number of petroglyphs on Judaculla Rock are impressive but the detail and intricacies of these figures can be difficult to appreciate when viewing the rock in person. This is because (unlike the contrasting colors on the "desert varnish" rocks in the Southwest) the carved and uncarved portions of Judaculla Rock are of a uniform color and do not provide the contrast that would be helpful in making out all the shapes. Excellent artists' reproductions of all the figures on Judaculla Rock are available in books and on related sites on the internet. These reproductions are well worth viewing before and after your visit to Judaculla Rock.

Judaculla Rock

The Bunker Brothers of Surry County

On April 13, 1843, the Bunker brothers each married a Yates sister—daughters of a tavernkeeper in Wilkesboro, North Carolina. That may not have been so unusual an event for the sparsely populated, agriculture oriented state of North Carolina, in the years before the Civil War. But the Bunkers were not your usual Wilkes County bachelors come a-courtin'. To begin with, the Bunkers were not originally from North Carolina and in fact had not spent much time in the state. The Bunkers were well educated and worldly. They spoke several languages. They had traveled extensively in America, Europe and Asia. They had met the Russian czar, the king of Siam and P.T. Barnum. And one other thing—the Bunker brothers were connected to each other by a thick fleshy ligament that joined them to each other at about their sternum. They were born that way and remained so connected for their entire lives. Chang and Eng Bunker were Siamese Twins—in fact, they were the original Siamese Twins.

Chang and Eng's path to North Carolina was a complicated one. The twins were born to a fisherman's family in 1811, on the Mekong River in present day Thailand. At an early age they were taken to the court of King Rama, the king of Siam, where they were ogled and examined. It is rumored that the king considered executing the children on the advice of some that associated their condition with bad omens for the kingdom. But in the mid-1820s, Chang and Eng, with their mother's permission, were taken aboard a ship that was sailing for America. They arrived in New York and were soon being exhibited to amazed throngs of paying customers. They proved to be a popular draw. As the shows and exhibitions—and their fame—increased, the twins added amazing synchronized acrobatic routines—somersaults and back flips—and shows of strength to their usual performance of simply being observed, and answering questions from the audience. Physically and mentally, Chang and Eng appeared to function in complete harmony—so much so that they were sometimes referred to as one person, i.e., Chang-Eng. They participated in grueling travels and countless performances in the United States and Europe, ultimately touring with P.T. Barnum. By the early 1840s Chang and Eng were world famous as the Siamese Twins.

So why Wilkesboro? Well, it does seem plausible that by 1843 the fisherman's sons from the Mekong tired of the cities, the travel, the crowds and the endless performances of their tours. By this time they had also attained some financial savvy and security. Something about the backwater of Wilkesboro in the 1840s must have appealed to Chang and Eng. Maybe they saw a chance for a more normal life—although if so, they must have been true visionaries. And then there were the Yates daughters, Adelaide and Sarah. It is known that romance first bloomed between Chang and Adelaide, and that only later did Eng and Sarah become interested in each other— either because of extreme good luck or maybe a clear sense of the inevitability of such a union given Chang and Adelaide's intentions. It is also said that there was initially substantial opposition to these unions from both the Yates parents and the community. But the double wedding took place in 1843, and the two couples moved into a farmhouse twelve miles northeast of Wilkesboro, in the Traphill Community.

Although they continued touring from time to time to supplement their incomes as farmers, Chang and Eng established themselves as solid members of the community. In the late 1840s and early 1850s they accumulated 1,000 acres of farmland on both sides of Stewarts Creek near White Plains in Surry County. They were slaveowners and they were active in the local Baptist church. They built separate houses on either side of Stewarts Creek, one house for Chang and Adelaide and the other for Eng and Sarah. By agreement, the twins would stay three days at one house and the next three days at the other house. And children began to accumulate. Chang and Adelaide had 10 kids, and Eng and Sarah had 11—21 kids between them.

The twin's famous harmony was apparently tested during their married life. Their wives would quarrel, the economy was unstable and the Civil War disrupted the region for many years. Union General George Stoneman's raiders invaded the area in 1865 and destroyed part of the Bunkers' homesteads. Chang also became more prone to abuse of alcohol while Eng remained more reserved and intellectual. Economic problems forced the twins to go on tour to Europe one last time in the late 1860s. Some speculate that the twins also went to

Europe to investigate the possibility of a surgery to separate them. Over the years many physicians had examined Chang and Eng with such an operation in mind but the decision to operate had never been made. Apparently this last investigation was also inconclusive. In 1870, on the ship returning the twins to the United States from Europe, Chang suffered a stroke, leaving him partially paralyzed on his right side.

The twins' understanding with the local Surry County doctor was that if either twin died the doctor would immediately sever the band that connected the two. But it was not to be. On the morning of 17 January 1874, Eng woke to find Chang cold beside him. Within hours, Eng too had died, the doctor not having been located in time. Later autopsies indicated that the band of flesh connecting Chang and Eng could have been severed without endangering either of their lives.

Chang and Eng Bunker donated land and helped build the White Plains Baptist Church, which is located south of Mt. Airy on old U.S. 601. Near the new brick church stands the old white clapboard church built in 1858 which the Bunkers attended. The twins are buried in the White Plains Baptist Church cemetery on the north side of old U.S. 601.

Eng and Sarah's house burned in 1956 and a new house was built on the site. This house site is at the northeast corner of Highway 601 and the interstate connector interchange—all just south of Mt. Airy. Stewarts Creek is nearby.

For more details about Chang and Eng Bunker see:
- The "Official Page Of Eng and Chang Bunker, The Original Siamese Twins," on the internet at
  http://engandchang.twinstuff.com/
- Judge Jesse Franklin Graves' *Life of Eng and Chang Bunker, The Original Siamese Twins,* (explaining that Chang and Eng assumed the name Bunker in honor of a New York family named Bunker with whom they had become close) at
  http://engandchang.twinstuff.com/graves_manuscript.htm
- Darin Strauss, *Chang and Eng, A Novel* (Plume Books, 2001)

The Silver Cup

My sister won every sailing regatta she ever entered. A regular corsair of the seas, she was, mateys. Her specialty was skippering a one-sail, dinghy like boat called an El Toro, around several buoys bobbing in the drink. To be truthful, the El Toro is a runty little boat—about eight feet long, with a squared off bow, and a shovel symbol on its sail. The shovel has something to do with the El Toro classification—the bull. If you are smirking to yourself about this connection, well ... you are probably right. Why else a shovel? But I digress. My sister and her El Toro sailed best in very light winds. The kind of weather when the boats just seem to sit there and the sails wrinkle and luff. Watching sailboat races from the shore is a tedious business at best, but watching a becalmed sailboat contest is a torturous affair. In any event, my sister had the foresight to retire early from the regatta scene and to save her streak from the disappointment that racing as an adult would surely have entailed. And retiring at the ripe old age of twelve ensured the unique status of her grand silver cup—the first place prize in the Governor's Cup Regatta of 1964.

In recent years, my mother has explained to me that she wanted her three kids to at least taste all the different treats on life's buffet line. So that is why, in the summer of 1962, she signed us all up for sailing lessons at some marina near the freeway in Berkeley, California, where we were staying for the summer . My mother says she had to wait to sign us up until we were all old enough for the lessons. In the summer of '62, my sister was ten and a half, my brother eight and a half, and I was seven. I remember two things about these sailing lessons. Just a foot or two below the surface of the water of the lake we sailed in were masses of horrible looking seaweed—probably, in my seven year old mind at least, the nasty home of eels and jellyfish and worse. The sight of these lurking weeds terrified me. The sailboats were always heeling over in the wind, with the waves lapping only inches from the gunwales. At all times I expected that the little sailboat would capsize and spill us into that water, into the weeds—to struggle to keep our feet from being swallowed by the morass, and probably to be pulled slowly under. I did not want to be out on that water in a boat that leaned over when the wind blew. I did not want to learn how to sail. I wanted to be safe on the shore. Or better yet climbing on Indian Rock in the Berkeley Hills near where we lived. And the other thing I remember is that all the boats in this sailing school were El Toros.

After that nautical summer of '62, our family moved to North Carolina, and bought a house on a lake in Chapel Hill. Naturally, we needed a boat. And I guess, naturally enough, my parents' thoughts turned toward those stubby little California boats. They bought first one wooden El Toro, and soon after a flashy red fiberglass El Toro. The first one was christened "Wee Three," and the second "Hurry Up Yawl." Quite witty if you are into sailing—which, alone in the family, I wasn't. In the winters of 1962-63 and 1963-64, I was into Duke basketball. Duke was quite good in those days. Those were the days of Jeff Mullins, Steve Vacendak. Jack Marin. Hack Tison. "UCLA, who are they?" we all yelled at Cameron Indoor. I did not care to go to Kerr Lake and Henderson Point to sail those stumpy boats. I didn't like water that wasn't in a swimming pool. Weeds, stumps, frogs—who knew what was down there. But we'd go, they'd sail and I'd wait on the shore, imagining my drives to the hoop and pitching rocks into the deep.

And then, one day, the adults got an idea. There were two other El Toro's in Durham, owned by two Duke professors (my father was also a Duke prof) as best I can recall. The El Toro sailors were regularly blown out of the waters of Kerr Lake by the classy, speedy Flying Scots, Lightnings, and the other larger boats with jibs and big sails. Those serious sailors were all training for the Governor's Cup Regatta held every year in North Carolina since 1957. One of the profs checked the rules and learned that if one could assemble three boats in a given class—three El Toros, for instance—the Governor's Cup Regatta would include a race for that class. And award a silver cup to the winner. Even if the three boats were stumpy little dinghy boats. Application was made and the entrants signed up. And it was made official. The 1964 Governor's Cup Regatta would include an El Toro competition.

Recently I've asked both my parents and my sister why my sister was chosen as the skipper of our family's entry in this race. My father simply said that he wasn't a sailor so it had to be Marjorie. As for Marjorie, it seems the issue had never occurred to her. She was the eldest and probably more of a sailor than anyone else in the family. And she was probably viewed by my father—and the other two Duke professors—as simply filling out the field so that the two adult sailors could fight it out for the cup. Interestingly from my sister's point of view, although my father swears there were only the three El Toros in the race, my sister recalls that there were five to seven boats competing.

I recall little about the race although I was certainly there—probably flipping rocks into the water. From my vantage point on the shore, I would have seen a bunch of sails, moving slowly, very slowly, in the distance. The lake water lapping on the red clay shore. My mother says she rubbed furniture wax on the hull of Marjorie's El Toro before the race. My sister remembered it as some sort of oil. But everyone agrees that by the time the El Toro division began its race, the wind had died down to a whisper. The boats barely moved. The sun bore down and the skippers pushed their tillers back and forth to try to generate some forward movement. The Toros floated toward the first buoy and very slowly the boat with the lightest skipper pulled in front. And that was the story of this race. The twelve year old girl was faster because she was lighter, and the Duke professors could only sit, mature and sweaty in their boats, hope for some swaying in the tall pines on the far shore of the lake (indicating the approach of some longed-for breeze), and watch as my sister proceeded them around the buoys and across the finish line.

I think I remember Marjorie walking up to the table to accept her award for finishing first—but our family has several black and white photographs of the event and I may just be remembering these photos. But the silver cup she won ... well, that cup is etched in my mind, even though I'm sure I haven't seen the cup in over twenty years. Six inches high, the cup is engraved with the words: "Governor's Cup Regatta 1964 1st Place." No one in the family had ever won such a prize. It was hugely admired. The cup was a family conversation piece and icon for years—heck, it still is even if we haven't seen it in awhile. For years it stayed on the shelf above the fireplace in our home. But Marjorie never competed in another regatta. Indeed the '64 Governor's Cup was her one and only regatta. Our two El Toros endured sporadic sailing, and lots of rowing and paddling, over the years until they finally fell into disuse, disrepair and final resting places on the shore. The cup eventually went with Marjorie when she left home for good after college. Maybe Marjorie used to handle the cup from time to time and remember the light breeze on her cheek as she bobbed away from the other boats. But maybe not. She and her husband drank champagne from the silver cup at her wedding in 1979. But then the cup must have gradually receded from her family's consciousness. There were jobs, kids, there was life. Marjorie says she's not sure where the cup is now. Packed away ... somewhere. I've asked her to find it.

And me? Even though I loved the cup, I continued to detest sailing—for many years. I stuck with the basketball though. I would play for hours on my backyard dirt court every day that I could. I'd play throughout the winter and my hands would become raw and cracked. Vacendak and Marin graduated and Duke hoops started slipping. But I stuck with it despite Duke's slide (Bob Verga became my man). And then, in the seventh grade—pay dirt. I was the tallest in my class, and my inside game was dominant. I was the star on the school team. In one game, we won 22-2, and I scored 12 of our points. Life was good. And then I stopped growing. By ninth grade I was a small forward, and by high school I was an average sized guard with my inside skills intact but no outside shot. None at all. This part of the story really goes no further—some enjoyable intramural play, I guess, but no more basketball glory.

But there was still glory to be found—in particular, a glorious early fall day in the late 1980s. On a whim, I accepted an invitation from some non-sailing friends from school to drive up to the dreaded Henderson Point on Kerr Lake—to go sailing. We rented a Sunfish. I told them I knew how to sail and I'd show them how. And sure enough, somehow I did know how to sail—knew exactly how to do it. It was a warm day and there was a steady, stiff breeze. I wheeled that boardboat all over the lake—to the far shore and back, tacking up the lake and blasting back down before the wind. I sailed like a master without even knowing that I knew how to do it. The passengers lying on the deck applauded my skill as they were soaked by the spray. It all just fell into place, on that golden, blustery afternoon.

The boat and I were one with the wind and the water. I was a sailor after all.

So when my sister tells me she has found it, I'll buy the champagne and meet her by the shore. My sister and I will take turns hoisting the icon and toasting. First to the breeze. Then to the waves. Then maybe the buoys, and even the El Toros themselves. Maybe a small one for Bob Verga. And we'll both drink deep from the silver cup.

Marjorie at the tiller of the El Toro

Marjorie and the Silver Cup

Hellespont Dreams: Cross-Training with Lord Byron

I've been running for thirty years. That's a lot of miles even if I only rarely did a 50 mile week. I did 10Ks in the 70s, marathons in the 80s, 5Ks in the 90s. I had fun, yes, I did. But I'm not getting any faster. Let's be more precise. I've gotten slower. Much slower. In the fall of 1999 I was back in training for the traditional seventeen mile mountain trail run that I usually do each fall, when I became seriously annoyed at how tough it had gotten running up this certain hill on our usual Sunday morning course. Between gasps for air I put it to my Sunday morning running partners huffing up the hill with me: "Why shouldn't I gracefully discontinue this running business in the new millennium?" Now this all happened early in a Sunday morning long run so they had plenty of time to formulate arguments and analysis and to talk me out of it. But they had nothin'. "Keep running because that's what you do and how could you not do it?"—pretty much sums up their response. So I kept running but I also kept thinking. I finished up the old millennium by skipping the seventeen mile trail run and realizing that I could expect some severe Y2K running problems.

The year 2000 is now one third over and my running buddies have not seen me at any of our running club's recent events. I am also the rare attendee at our Sunday morning long runs—and if I show up I'm always at the back of the pack and, after a mile or so, I drop further back and eventually disappear from the sight and consciousness of the regular group. Ah, but don't think I haven't been training. Because I have. And now I have a date with destiny and a dead poet.

So, its now a little after 10:00 a.m., on May 3, 2000. It is one hundred and ninety years to the day, to the hour, and even to the minute that George Gordon Byron, Lord Byron to his admirers and imitators, waded into the waters of the Dardanelles Straits, to swim from Europe to Asia. I also am dangling my toes in the same murky roiling waters, Speedo and goggles on, ready to follow in Byron's footsteps—well, okay, I guess more in his wake, actually. I empty my mind of twentieth century cares, silently recite my favorite Byronic stanzas, and slip into the chilly water and begin to stroke for Asia— visible as only a dark margin along the horizon where the water seems to merge into the haze. Well, okay, to be precise, I can't actually see

Asia and the water isn't actually all that chilly or murky and roiling. And its only really the same water in the sense that all water is water, after all. I'm actually in the pool at the Pullen Aquatic Center in Raleigh, North Carolina, and I'm making my first flip turn. There. I push off, glide and come up for air. One lap down, 140 or so to go. You know I don't have the time or the money to go to Turkey and do the actual swim. I've got no hereditary estate like Lord Byron did in 1810. I've got a day job. And besides the Dardanelles is supposed to be so horribly polluted these days that no sane distance swimmer would be caught dead in it (so to speak). But still, it's the right time (if not the right place), it's the right distance, it's water, and I'm in the proper Byronic frame of mind. So let's call it the Hellespont, okay? Work with me here. I do another flip turn, much as Byron himself would have done had he trained in an Olympic size pool like Pullen. I can sense the great continent of Asia drawing nearer. I'm making good time.

Byron considered himself an accomplished distance swimmer. In 1810, at the age of twenty-two and on his way to Istanbul (then Constantinople), Byron challenged Lieutenant Ekenhead (an officer of the ship on which Byron sailed) to a swimming race across the Hellespont (now called the Dardanelles). Swimming the Hellespont also appealed to Byron because of the Greek legend that Leander swam the Hellespont nightly to meet his lover, Hero, a priestess of Aphrodite who lived on the other side and had to be visited in secret (Note: after each of these assignations Leander presumably re-entered the chilly water and swam back to Asia; like Byron, I concluded that a one-way Hellespont paddle was sufficient challenge; according to legend Leander did ultimately stop his nightly round trip swims after he drowned one night on one of the legs). Byron finished the swim in an hour and ten minutes but Ekenhead finished in an hour and five minutes. On the day of the swim, Byron wrote that the distance "is not above a mile but the current renders it hazardous, so much so, that I doubt whether Leander's conjugal powers must not have been exhausted in his passage to Paradise." Byron was very proud of his swim (he is said to have never allowed his friends or the public to forget about the accomplishment), and the achievement apparently grew in his mind as he recollected it. Several weeks after the event

Byron wrote: "The whole distance Ekenhead and myself swum was more than four miles the current very strong and cold, some large fish near us when half across, we were not fatigued but a little chilled. Did it with difficulty." So was it a mile or a four mile swim?

I'm about a mile into my swim across the Hellespont. I think I'm about half way so I look for the large fish but see none. A couple of large lap swimmers slide by in the adjoining lanes. From the looks of them, they will never make it to Asia. I keep stroking, having hit my stride . . . uh, I mean my . . . uh, stroke. I'm swimming freestyle—the old Australian crawl—and in an hour and ten minutes I expect to cover about two miles or about 140 laps in the pool. I figure that must be about the distance Byron covered—not including the distance he was pushed by the current. In 1810 there was no Australian crawl. Byron and Ekenhead swam breaststroke all the way across the Hellespont. They may have been strong breaststrokers but they did not breaststroke four miles in 70 and 65 minutes respectively. I'm thinking my freestyle is about as fast as Byron's breaststroke. I'm doing flip turns after all and freestyle is acknowledged to be the fastest of the strokes. But I don't want to stop short of the Asian shore and drown. So I guess I'd better swim at least an hour and fifteen minutes and maybe do two and a quarter miles. I'm still looking for those big fish.

At a mile and a half I feel pretty good and decide to increase the effort and try to catch Lieutenant Ekenhead. I can make out his head bobbing about a hundred yards ahead of me. But the effort proves a bit much. I decide that it would fit the spirit of the swim if I do a few laps of breaststroke. Then I decide that Byron probably also threw in some backstroke so he could meditate on the sky and savor the moment. So I do some backstroke and watch the ceiling of Pullen Aquatic Center move stately by. I'm slowing down but I switch back to freestyle and make two miles at an hour and fifteen minutes. Five minutes later I touch sand with the fingers of my stroking hand. Asia, at last! I gather my legs up under me and stand in the thigh deep water. Pulling my goggles up to my forehead, I walk stiffly out of the Hellespont and onto the Turkish shore. As the waters of the Bosporus drip off my body, I turn to look back across the strait toward Europe. As I gaze across the murky, roiling water, I dip my head slightly in mute salute to and acknowledgment of Leander, Lieutenant Ekenhead and, of course, Lord Byron himself. Then I turn to the lifeguard who is looking at me curiously. There are no other lap swimmers left in the pool. "All done," I say, smiling and realizing that the lifeguard hadn't been on the stand when I'd started swimming so he couldn't know where I'd started from or just how far I'd come. And I won't tell him. I turn away, put my arm on Byron's shoulder and walk with him to the showers.

So now that I've swum the Hellespont, what's next? Disneyworld? Reenacting the 1808 footrace between Mountain Man John Colter and warriors of the Blackfoot tribe where the braves captured Colter, stripped him naked, then gave him a head start and chased him for 200 miles until he reached the safety of Fort Raymond? Hmmmm. You know, I'm thinking that it would be rather Byronic to start training for one of our upcoming local 5K running races. I may need to show up at the next Sunday morning run and ask my buds what they think? Lord B. would have understood, I think. As he said, or might have said, in one of his poems:

So we'll go no more a-swimming,

So far across the 'pont,

Though these arms be used to stroking,

On that legendary jaunt.

But my Speedo grows a-faded,

and my pool pass expires in June.

So it's back to 5K races,

Throw my goggles at the moon.

Faith Rock

It's very early April and the bare trees of the hardwood forests on either side of Highway 64 show not the slightest tinge of green. But near the edge of the road's right of way there is the occasional neon purple glow lighting up the browns and grays of the deeper woods. The redbud is abloom. The Piedmont spring is foretold. Winter is over.

I've passed through Siler City and Ramseur. Soon I drive over the Deep River. I'm now in the heart of what once was the country of Colonel David Fanning. David Fanning—the bald headed, silk skull-capped, Loyalist officer in the King's militia who shot, hung and murdered many patriots in Randolph and Chatham County in the early 1780s during the Revolutionary War. Fanning was a grim, but memorable, kind of guy. One historian, Samuel A. Ashe, has commented that Fanning "was one of the boldest men, most fertile in expedients, and quick in execution, that ever lived in North Carolina. Had he been on the Whig [Patriot] side, his fame would have been more enduring than that of any other partisan officer whose memory is now so dear to all patriots." Even after Cornwallis' surrender at Yorktown in October of 1781, Fanning continued to lead a deadly backcountry civil war against the patriots until he fled to South Carolina in April of 1782. It should be mentioned that when, after the war, the North Carolina legislature passed the generous Act of Pardon and Oblivion, which granted clemency to most former Loyalists in the state, three individuals were specifically exempted because of the egregious nature of their depredations during the war. One of them was David Fanning.

Fanning's exploits included an invasion of the town of Pittsboro where he took 53 prisoners which included "virtually every important citizen in Chatham County," and later in 1781, a raid on Hillsborough where he took 200 prisoners including Governor Thomas Burke. Fanning also led the Tory band which attacked Philip Alston and his family at the House in the Horseshoe in Moore County in August of 1781. Bullet holes from this attack can still be seen in the walls of the house which is now a state historic site. Another famous Fanning story involves a patriot's lucky escape from Fanning's rough justice by use of Fanning's favorite horse, a great exposed rock jutting into the Deep River and the river itself. It also happened in that busy year of 1781.

Across the Deep River and heading west on Highway 64, I turn right onto Faith Rock Road. This road ends at Andrew Hunter Road (State Road 2235). I turn right and ride down the short hill to the Andrew Hunter Bridge over the Deep River. A few hundred yards further along this road I pass an old abandoned factory on the right and just past the factory I see the small parking area (also on the right) and the Andrew Hunter Pedestrian Bridge. I park and walk across the narrow pedestrian bridge glancing upriver—but I see no rock, just a broad, deep river. Once across the bridge there is a sign that says "Faith Rock" and an arrow pointing at a trail that heads off to the right along the river. I follow the trail.

Andrew Hunter was an outspoken Randolph County Patriot. David Fanning had sworn to capture and execute Hunter. One morning Fanning and his men succeeded in capturing Hunter. They determined they would hang him after they had their lunch. Somehow, during their repast, Hunter was able to free himself and jump upon David Fanning's favorite horse, called Bay Doe, and to ride off. Fanning, known as a superb horseman who treasured his mounts, is reputed to have ordered his men to give chase and "kill the rascal but spare the mare." This order may have caused Fanning's men to hold their fire, or at least hesitate before firing at Hunter, for fear of wounding the horse. So Hunter galloped away toward the Deep River. He was trying to reach one of the several fords that would let him cross the Deep River and escape. But each ford he came to was guarded by Fanning's men. Desperate and running out of time, Hunter rode up a bluff overlooking the river. He could see that between him and the river below was an exposed rock face that sloped down to the river at an angle of some sixty degrees or so. Facing capture if he lingered, Hunter spurred his horse down the steep rock slope and plunged into the murky waters. Bobbing to the surface, Hunter and Bay Doe then floated down river. Fanning's Tories apparently were so surprised and impressed that they fired no shots at the swimmers. Reportedly one of Fanning's men exclaimed, "If he has faith enough to try to escape that way we will not shoot again."

I walk along the river bank until I see the huge exposed rock, steeply sloping up to my left, that would indeed provide a clear path from the high river bank down to the river—if you had the nerve and the skill to urge a horse down its terrifying expanse, hooves skittering on the bare rock, knowing there would be no place to stop short of the cold waters of the Deep River. It must have been a mighty splash. And an impressive feat to watch. Certainly no one was going to give chase, at least not by following the route Hunter had taken. Col. Fanning must have been furious. He lost his prized horse and pistols and valuable papers that had been in the horse's saddle bags. Presumably Hunter rode Bay Doe up out of the Deep River some ways downstream and successfully avoided Fanning and his men to return to his home. And although it isn't really clear that Andrew Hunter had any other choice, the big rock has been called "Faith Rock" ever since—a monument to troubled times and one man's remarkable luck and spirit.

I hike back along the river and cross the pedestrian bridge. I drive back out to Highway 64 and then head northwest across Randolph and Davidson Counties. I cross the Yadkin at the famous Shallow Ford and follow the ever-expanding improvements to Highway 421 toward my destination for the night, Boone. Once through Deep Gap, high on the Blue Ridge plateau, I see no more neon glows in the forest's edge. It is winter once more.

For more on David Fanning and related matters, see:

- David Fanning, *The Narrative of Col. David Fanning*, published December 1861; Fanning was born in Virginia in 1755; his father drowned in the Deep River before David was born and his mother died when he was nine; Fanning and his sister were raised in Wake County by a guardian; Fanning suffered from an "offensive scalp disease," called "scald head" or "tetter worm" which left him bald and caused him to usually wear a silk skull cap; Fanning was deported to Canada in 1784 and he died March 14, 1825 in Nova Scotia.
- Daniel W. Barefoot, *Touring North Carolina's Revolutionary War Sites*, (John F. Blair, Publisher, 1998) for information about North Carolina during the Revolutionary War, David Fanning, Faith Rock and the House in the Horseshoe.

Faith Rock above the Deep River

Pausing Beside the King's Highway

The trees covering the hills outside Beijing, China, are mostly small and scraggly—a result of many centuries of timbering. But high up in the hills is an ancient Buddhist temple which is even older than those early tree-cutting endeavors. Surrounding the shrines and courtyards of the temple are towering, majestic trees that have been protected from the saw by the local populace's respect for the temple. One of the trees is taller and larger than the rest. It is huge and said to be well over a thousand years old. The tree was visited regularly by Chinese royalty on day trips from their palace in Beijing's Forbidden City. This grandest of trees also has a name: The Emperor of Trees.

You gotta love trees that have survived for hundreds of years and grown huge while anchored to one lucky spot on this earth. And more impressive still are the trees so grand that they have been given names. Like the Angel Oak on John's Island, near Charleston, South Carolina. It's a massive live oak, 65 feet tall, 25.5 feet in circumference and providing 17,000 square feet of shade. Some of its large branches extend over 80 feet from the trunk. The tree is thought to be 1,400 years old. I visited the Angel Oak in the late 1980s or early 1990s and remember having to look through gaps in a fence that surrounded the tree. At that time the tree was privately owned and had been closed to the public. But I understand that the tree is now accessible in the City of Charleston's Angel Oak Park. I've also visited Oregon's largest tree. Near Seaside, Oregon, just off Highway 26, is the largest Sitka Spruce in the continental United States. 216 feet tall and 750 years old, this tree is very, very big. How it escaped the loggers' clearcuts that dominate coastal Oregon is a wonder. And of course most North Carolinians are familiar with the Davie Poplar—the grand tree in the center of McCorkle Place near the Old Well on the campus of the University of North Carolina in Chapel Hill. Tradition holds that in 1792 William R. Davie stood beneath the tree (or thrust a poplar branch into the ground) and declared that that would be the site for the new University. Although the tree is now full of cement and held up by cables, its progeny, a Davie Poplar Jr., was planted nearby in 1918. And then there is another special North Carolina tree which endures only several feet from the cars and trucks whizzing by on the blacktop of a highway in Pender County.

In the late 1700s the road between New Bern and Wilmington was sandy and primitive. It was known as the King's Highway and it would eventually be improved and renamed as Highway 17. One who traveled the King's Highway in 1791 described the route as follows: "The whole Road from Newbern to Wilmington (except in a few places of small extent) passes through the most barren country I ever beheld; especially in the parts nearest to the latter; which is no other than a bed of white sand." The traveler was the President of the United States, George Washington. Washington was in the midst of his Southern Tour—a loop through Virginia, North and South Carolina, and Georgia, designed to help the President discover "the temper and disposition of the inhabitants towards the new government." Washington would visit New Bern, Wilmington, Charleston, Savannah, Augusta, Columbia, Charlotte, Salisbury, and Salem, on this nineteen hundred mile tour. For the most part, Washington traveled in a "magnificent" white carriage pulled by four horses, with designs of the four seasons and the Washington coat of arms painted on the sides. Washington called it his "chariot." It is said that upon approaching a town, where the inhabitants would generally turn out en masse to cheer the President, the 59 year old Washington would exit the carriage and mount his handsome stallion, Prescott, for a carefully choreographed entrance. Washington's traveling companions numbered only five other persons, although the locals would often accompany him for lengthy stretches after leaving a settlement.

On April 23 or 24, 1791, Washington was nearing Wilmington, North Carolina, when his entourage stopped to rest in the shade of a large oak that stood next to the King's Highway. That tree, the Washington Oak, still shades travelers on Highway 17—although today's travelers are all zipping by in automobiles at thirty-five miles an hour and care little for the shade. The Washington Oak is not well marked but if you are looking carefully it's size and immense branches are easy to spot. It is just south of Hampstead on the west side of Highway 17, several hundred yards south of the intersection with State Road 210. If you choose to rest under the tree you will know it is the right tree if you see a stone marker next to the massive trunk placed by the D.A.R. with a plaque that commemorates Washington's stop to rest beneath the branches of the Washington Oak.

Carolina Journeys

The Washington Oak in Pender County

The Bigfoot Chapel Hill Hash

As previously noted, the Hash House Harrier phenomenon arrived in North Carolina in the early 1980s. Hashers get together on a regular basis to choose one of their kind to be a "hare." This hare will lay an intermittent trail marked with dollops of flour, and the remaining hashers—or "hounds"— attempt to follow this trail. At the end of the hash there is always food and drink—the famous hash après. Hashers are known by their quirky yet somehow appropriate hash nicknames, names they earn after proving what they are made of over the course of several hashes. This report, known as the Hash Trash in hashing circles, concerns a late summer hash course laid in Chapel Hill by a Scandinavian hash enthusiast with a charming accent and a couple of large pedal parts.

No, Bigfoot ain't from around here, folks. He mysteriously says he comes from somewhere in the west—Burlington or Graham or some other alluring and unpronounceable name. But I think he may really be from somewhere down east—I'm thinking Burgaw or Bug Hill. There are those stories the old folks tell about the legendary Bigfoot who stumbles out of the swamps of Robeson and Columbus Counties on misty winter nights to terrorize the locals. Coincidence, you say? Maybe so. Maybe so. All I know is that the Danish word for monsterhashlayer sounds awfully like "beegfute." What's that all about?

So Bigfoot says to himself—I'll lay this hash in Chapel Hill from my daughter's pad and I won't leave anything out except Finley Golf Course and the Dean Dome. Maybe a little voice echoed in Bigfoot's head saying: "And you'll do this in about five miles and about an hour, right?" But if he heard that voice, Bigfoot ignored it. Can't leave out those woodsy bike trails off Estes Extension or the student houses and keg parties in residential Carrboro, Bigfoot's fevered mind reasoned. And then there is Carr Mill Mall to be run through and the hash has to run right down main street in Carrboro. Then head on over to Chapel Hill and stop for a beer break at one of the pubs on Franklin Street. Okay, Mr. Foot, all that sounds good— and then we jog home, right? Well, what about running down Franklin Street for the sheer glory of it, Bigfoot whispers out loud. He's nodding his head as he talks to himself. And the opportunity to run all the way through NCNB Plaza (and look at ourselves in the hallway mirrors) and then burst through the glass doors leading to Franklin Street again—passing the line of chanting Hare Krishnas and stopping at the cross walk to sing a contrasting tune of past hashing heroism (possibly amusing said Hare Krishnas, but possibly not). Then on to those must-see symbols of the oldest state university—the Davie Poplar and the Old Well—where the photographer was taking pictures of the wedding couple with the best man looking on, and the dollops of flour led right between the newlyweds (I was sure there was a dollop squarely underneath the bride but we never did find out for sure). The best man smiled at our taunts but the bride was surly. Balmy looked at Feces and Feces said, "On-on!" Okay, that was fun, Mr. Foot, but when do we get to go home? Got to do the arboretum, says the evil Foot, then by Bill Friday's house and past the old library, and (of course!) down the hill through the woods back to the Bolin Creek bike trail and the Bolin Creek express back to the après. Was it eight miles? Was it two hours? Was I a beat 'Wurm? Certainly t'was. Oh, did I fail to mention the hare's insistence on running part of the way with blue ribbons attaching staff and distaff hashers? Wow, now there's a tradition that's bound to catch on at hashes!

So, Bigfoot did finally get us all back and Mrs. 'Foot put on quite a spread at the Foothouse. We caroused on the back porch as the light faded from the sky. We razzed Micro for being porcine and lazy. We razzed Alden for not having a decent hash name. We razzed people we didn't even know and razzed people who weren't even there. We razzed the young friends of Bigfoot's daughter, one of whom asked if Endangered Feces was Feces' real name. We assured him it was. We razzed Bigfoot until the air grew misty and a cloud covered the moon and we remembered the old folks' tales and hastily exited Bigfoot's lair, looking over our shoulders with hearts beating rapidly, until we reached the safety of our cars and locked the doors.

For more information on hashing see:
- Tar Heel Hash House Harrier web page:
  http://pages.zdnet.com/commentateur/tarheelhashers
- Sir Walter Hash House Harrier web page:
  http://www.swh3.com/

Going After Elisha

The year was 1794. French botanist Andre Michaux had just climbed to the top of Grandfather Mountain. He looked around, didn't see anything higher, and concluded he had reached the highest point in North America. If you've been to the top of one of the rocky crags along Grandfather's ridge, you can understand what Michaux was thinking. It does look like it could—and should—be the highest thing for many miles around. So, according to the *WPA Guide*, when Michaux reached the summit and decided he was on the high point on the continent, he "triumphantly sang the Marseillaise." His spirit and obvious elation are both admirable and contagious—but Andre, of course, was wrong. Wonderful mountain though it is Grandfather isn't the tallest mountain in North America. It isn't the tallest mountain east of the Mississippi or even in the Southern Appalachians. Grandfather isn't even in the top forty of the highest peaks in the Southern Appalachians. On the other hand, in the first half of the nineteenth century, many Americans—particularly those in the north—believed that New Hampshire's Mount Washington was the highest mountain east of the Mississippi. At 6,288 feet and high above the tree line, Mount Washington's summit does certainly look to be a fitting candidate. But some who had seen them thought that the highest peak in the east might be in the Black Mountains northeast of Asheville, North Carolina. One of these was a professor at the University of North Carolina, Dr. Elisha Mitchell.

Elisha Mitchell left his native Connecticut to become a professor at the university in Chapel Hill in 1818 when he was 25 years old. At that time the university had only 92 students and three professors. Probably of necessity, Mitchell's interests were many and varied. He taught a variety of subjects at the university from 1818 until his death in 1857. In 1828 he traveled to the west and climbed Grandfather Mountain. At that time he wrote that based on his observations, he suspected the Black Mountains and Roan Mountain were higher than Grandfather. In 1835, to test his hypothesis, Mitchell paid his first visit to the fishhook-shaped ridge known as the Black Mountains.

I drive up the Blue Ridge Parkway from Asheville, past Craggy Gardens, and I park my car at Balsam Gap, at Milepost 359.8. I follow the trail that starts at the Gap northwards along the ridgeline. To the east and far below is the Cane River Valley. Across the valley, and looming vast and dark is a huge ridgeline with several protuberances marking ridgetops. I cross over a high point on the ridge known as Little Butt and finally pass near the highest point on this ridgeline, known as Big Butt. In 1835 Big Butt was called Yeates Knob. In that year, guided by a local man named William Wilson, Dr. Mitchell climbed Yeates Knob (now Big Butt) and stared (as I'm now staring) at the dark, huge ridge to the east. Mitchell knew the various knobs on that ridge were higher than where he was standing but that it was "a matter of considerable difficulty" to determine which of the knobs was the highest because he was unable "to determine how much of the apparent elevation of one, amongst a number, is due to its nearness, and how much to height." I see the same knobs and have the same difficulty in judging their respective heights. To the left, or north, there is Cattail Peak, Balsam Cone, and the Big Tom/Mt. Craig complex. In the middle is Mount Mitchell. To the south is the Mt. Gibbes/Clingman's Peak/Potato Knob complex. They all look about the same height. You just can't tell which is highest by looking. You gotta go there. And take your barometer.

Although we don't know for sure, it seems likely that the next day Dr. Mitchell and William Wilson climbed the Mt. Gibbes/Clingman's Peak/Potato Knob complex, thinking that it was the highest point—although it is possible that they summited Mt. Mitchell itself. Dr. Mitchell's own accounts of this hike are not clear—and are even somewhat inconsistent. Again, confusion is certainly understandable. Mt. Gibbes is 6,571 feet high. Mt. Mitchell is 6,684 feet. Not a huge difference for an eyeball comparison—or even for a man with a barometer, which Mitchell had. Whichever mountain Mitchell had climbed, he used his barometer to measure its elevation at 6,476 feet above sea level—higher than Mount Washington. That fall Mitchell first published his conclusion that the highest peak in the eastern United States was in the Black Mountains of North Carolina. The news started to spread.

Mitchell had determined that the Black Mountains had peaks higher than Mount Washington—but had he determined the highest peak? Possibly harboring doubts as to whether he had actually climbed the highest knob in the Black Mountains in 1835, Mitchell returned to the Blacks in 1938 and in 1944. Both times he again climbed the Mt. Gibbes/Clingman's Peak/Potato Knob complex and took his barometric measurements. Apparently he was satisfied that he had found and climbed the highest peak—interestingly, the peak in the Mt. Gibbes/Clingman's Peak/Potato Knob complex now known as Clingman's Peak (the one that presently has all the antennas on it) was, in Mitchell's day, known as Mount Mitchell. The future Mount Mitchell, three miles to the north of the former Mount Mitchell, would soon come to be known as Clingman's Peak. Talk about your irony. But this is just the beginning.

Thomas L. Clingman was a North Carolina native who attended the University of North Carolina and was a student of Dr. Elisha Mitchell. Clingman was valedictorian of the graduating class of 1832. Clingman soon became a lawyer and shortly afterwards was elected to the state legislature—in later years he would serve in the U.S. House of Representatives and the U.S. Senate. But on September 8, 1855, Clingman was to be found in the Black Mountains following in Dr. Mitchell's footsteps. Clingman climbed to the top of Clingman's Peak (then known as Mount Mitchell as noted above) where he took a barometric reading. He then continued on along the ridge, heading north, until he summited present-day Mount Mitchell where he took another barometric reading. This reading indicated a pressure .19 inches less than Clingman's Peak—meaning he was at a higher altitude. Clingman thus became convinced that the present-day Mount Mitchell was actually the highest peak in the Blacks. Clingman also published his measurements and his conclusions. This publication had two results: (1) present-day Mount Mitchell began to be called Clingman's Peak; and (2) many began to doubt whether Elisha Mitchell had actually summited the highest peak in his 1835, 1838 and 1844 visits to the region.

Dr. Mitchell apparently took offense at Clingman's claims and in 1856 and 1857 both Mitchell and Clingman published additional articles explaining their positions and attacking the other's credibility. It became personal and public—at one point Mitchell accused Clingman of "injustice" and "wickedness." Yet in one article, Mitchell did admit that it was not until his 1844 trip that he actually climbed the highest point, that is the present-day Mount Mitchell. This admission raised more questions than it answered, however, about what mountains Mitchell had actually climbed in each of his trips. And Clingman still argued Mitchell hadn't climbed the right mountain even in 1944. Clingman believed that his own 1855 ascent of the high peak was the first.

It was in this context that the 64 year old Dr. Elisha Mitchell left Chapel Hill in the summer of 1857 for yet another sojourn to the Black Mountains—to take more measurements and maybe to clarify in his own mind just what peaks he had actually climbed. On Saturday, 27 June 1857, at about 2:30 p.m., Mitchell left the "Mountain House," a cabin south of Mt. Gibbes, and started hiking by himself to the north. No one is certain of his precise purpose but it is believed he intended to walk to certain settlements on the upper stretches of the Cane River where he would spend the night. To get there he would walk along the western side of the Black Mountains and then turn to the west and hike directly down the slope to the Cane River. But Mitchell never showed up at the Cane River cabins. And Mitchell did not return to the Mountain House on the next Monday as he had informed others that he would. When several more days passed without any sight or sign of Mitchell, two search parties were organized—one led by future North Carolina governor Zebulon Vance and the other led by local guide and bearhunter Big Tom Wilson.

The search parties found little in the area south of Mount Mitchell. But on Tuesday, July 7, 1857, in a spot a quarter mile or so to the west or northwest of Mount Mitchell, Big Tom Wilson found a couple of boot prints heading down Little Piney Ridge toward the Cane River Valley. The tracks led to a creek and then followed the creek downstream. Big Tom Wilson knew the area well. He knew that the stream soon tumbled down a forty foot waterfall. It is reported that Big Tom Wilson suggested to his fellow searchers that they would find Mitchell at the bottom of the falls. When the searchers reached the falls, Big Tom clambered down to the circular pool at the base of the falls. He first saw Mitchell's hat and then he found Mitchell's body submerged in the pool, pinned underneath a large log that must have been swept over the falls. "Here he is. Poor old fellow," Big Tom called up to his companions. They pulled Mitchell's body from the cold waters. In his pocket they found Mitchell's watch with the hands stopped at the likely time of the fall: 8:19 p.m., Saturday, July 27, 1857. It seems likely that Dr. Mitchell had been in a hurry to get to the Cane River cabins as he headed down the mountain in the gathering darkness on that Saturday night. He knew that the little creek would flow into the Cane River so he could follow the creek to his destination. But walking in or near the creek's edge, it was his misfortune to miscalculate or slip and loose his balance and fall into the water too close to the falls.

Dr. Mitchell's body was carried to Asheville and buried there on July 10, 1857. But greater forces were afoot. Various supporters of Mitchell—and detractors of Clingman—clamored for Mitchell to be buried on the Black Mountains' high point. The dramatic circumstances of Dr. Mitchell's death proved compelling and Mitchell's body was exhumed and reburied in June of 1858 on the very summit of Mount Mitchell—where he still lies today.

Maps of the likely route of Dr. Mitchell's last hike show him skirting the summits of the Mt. Gibbes/Clingman's Peak/Potato Knob complex on the western side and then heading down slope to Stepp's Gap—site of the present-day ranger station. Mitchell then started up the trail to Mount Mitchell—again also passing just below this summit on the western side. Once Mount Craig was directly to his east, Mitchell turned to the west and headed down slope toward the Cane River Valley. A steep slope leads you down to Mitchell Creek and if you head downstream you'll soon arrive at the top of Mitchell Falls. Be careful. You will recognize the deep circular pool at the bottom of the falls from the old pictures.

Dr. Elisha Mitchell lost the battle of his last hike but he clearly won the war with Thomas Clingman. With his dramatic death and reburial on the summit of the highest peak, there was little protest when the mountain was officially named Mount Mitchell, and it was soon forgotten that the peak was once known as Clingman's Peak. Clingman persisted in his explorations, however, and in 1858 he announced that a grand mountain in the Smokies, on the border of

North Carolina and Tennessee, was actually higher than Mount Mitchell. And he was close—and his championing of this mountain was enough to gain him naming rights. But Clingman's Dome tops out at only 6,643 feet above sea-level—forty-one feet under Mount Mitchell's triumphant 6,684 standard. Elisha Mitchell remains at the top. And Dr. Mitchell's silver-plated pocket watch—still stopped at the time of 8:19—what became of it? You can see the watch on the campus of the University of North Carolina in Chapel Hill in the school's North Carolina Collection housed in Wilson Library.

Postscript: If you make it to the top of Mount Mitchell, to climb the observation tower and visit Elisha Mitchell's grave, you should also visit an interesting site down a side trail a hundred feet or so below the summit. Not far down this trail is an expanse of exposed rock that protrudes horizontally toward the trail's edge—a natural shelter from the elements. This protected spot was a convenient and well-used campsite for those visiting the summit of Mount Mitchell. It went by various names: the "Shelving Rock," the "Sleeping Rock," even the "Black Mountain Hotel." When I visited the site the rangers said it was simply called "Camp Rock." It was a popular site in the 1850s and after the Civil War, as tourism to the highest point in the east increased. During the War it was used by deserters from the Confederate Army who were hiding from capture by the feared Home Guard. I'd like to tell you that Dr. Mitchell slept there or at least visited the spot—but I have not discovered any such report or reference in my research. Alas. Still well worth a visit though.

For more details about the Mitchell/Clingman controversy see:
- S. Kent Schwarzkopf, *A History of Mt. Mitchell and the Black Mountains* (N.C. Division of Archives and History, 1985)
- Timothy Silver, *Mount Mitchell and the Black Mountains* (University of North Carolina Press, 2003)

The Dismal Swamp Canal

Straddling the North Carolina-Virginia border far to the east of Interstate 95 is a "vast body of dirt and nastiness"—at least as described by the sardonic William Byrd in 1728. But Byrd was biased. He was angry because he had the job of surveying the dividing line between North Carolina and Virginia which ran directly through this "vast body" of wilderness, also known as the Great Dismal Swamp. It must have been a brutish, nasty and biasing job. The Swamp is indeed big, about 600 square miles—but in Colonial times the Swamp was much bigger, probably over 2,000 square miles. In the heart of the Swamp are the dark, shallow, tannin-stained waters of Lake Drummond—a 3,000 acre natural lake "discovered" by the first governor of North Carolina, William Drummond, in 1665. The Swamp's western edge is defined by the Nansemond Escarpment—a plateau which marks the original and ancient coastline of this area. The eastern edge of the Swamp is marked by Highway 17 and a 22 mile long man-made ditch known as the Dismal Swamp Canal. Since 1805, this canal has connected the Chesapeake Bay with the Albemarle Sound. It is the oldest continually operating canal in the United States.

William Byrd, the surveyor of the dividing line between North Carolina and Virginia, was the first to muse about draining the Dismal Swamp and building a canal through the swamp to connect the Chesapeake and the Albemarle. Years later the young entrepreneur George Washington became interested in the project. Washington toured and surveyed the swamp several times beginning in 1763, and he formed a company to undertake the projects. Although the company ultimately folded, it did successfully complete one drainage project—the "Washington Ditch," which runs northwest from Lake Drummond to the Nansemond Escarpment. This ditch still exists today. Talk of the canal began again in 1784 and counted Thomas Jefferson, James Madison and Patrick Henry among its supporters. The Virginia legislature passed supporting legislation in 1790. The actual digging began in 1793.

As can be imagined, the work moved slowly. The canal was dug entirely by hand, mostly with slave-labor, in trying conditions. Money was tight and labor always in seeming short supply. The digging began at both ends of the proposed canal—the Pasquotank River on the southern end and the Deep Creek tributary of the Elizabeth River on the northern end. The ditches were finally joined in 1805 but long stretches of the canal had not been dug to the required depth and width—the canal was described as "little more than a muddy ditch." It was also realized that the canal would need to be filled with water from Lake Drummond, so a three and a half mile long "Feeder Ditch" was dug in 1812 connecting the two. A series of locks were also built and deepening of the canal continued. Finally, on June 11, 1814 the first ship completed the passage from the Albemarle to the Chesapeake along the Dismal Swamp Canal. It was the first of many.

The Canal was a huge success from 1814 until the Civil War years. The British had blockaded the mouth of Chesapeake Bay during the War of 1812 proving the value of a protected intercoastal waterway to continued shipping in times of war. And the Canal was convenient and well-maintained. Commercial traffic—and so toll receipts—grew spectacularly in the first half of the 19th century. And a distinctive social life developed for Canal travelers. Next to the Canal, straddling the North Carolina-Virginia border, an impressive hotel was built in 1829, first known as the Lake Drummond Hotel and later as the Half-Way House. Because of its location (i.e., not easily accessible and close to the limits of each state's legal jurisdiction), the Half-Way House became a popular destination for gambling, duels, elopements and trysts. One legend reports that Edgar Allen Poe wrote his ominous poem, "The Raven," while a guest at the Half-Way House.

The Canal's very success led, however, to its undoing. In 1859, a new canal, the Albemarle and Chesapeake Canal, was opened that also connected the Chesapeake and the Albemarle. Because this canal was shorter and had fewer locks, it began to siphon off customers of the Dismal Swamp Canal. And the outbreak of the Civil War made the Canal a strategic target for both sides. In the spring of 1862, after defeating the Confederate's small naval flotilla and occupying Elizabeth City, Federal troops turned their attention to the Dismal Swamp Canal—to either capture or destroy the canal's locks located at South Mill. The Union generals were concerned that the Confederacy might use the Canal to send Virginia ironclads down to the North Carolina waterways. So the Federals marched on South

Mill with two thousand men. The Confederates had only several hundred men but they were entrenched in defensive earthworks and repulsed the Federals in a five hour battle marked by several frontal assaults and use of "roasted ditch" defense. The Canal remained in Confederate hands for a time, only to fall when the Federals captured Norfolk.

After the Civil War the Dismal Swamp Canal fell into major disrepair and the bulk of commercial traffic shifted to the Albemarle and Chesapeake Canal. Private investors funded some major improvements for the Canal in the late 1890s, resulting in a modest return to prosperity at the turn of the century. But this recovery was short-lived. In 1911 the federal government purchased the Albemarle and Chesapeake Canal, making it both federally maintained and toll-free—and ending the Dismal Swamp Canal's practical competitiveness. The federal government finally bought the Dismal Swamp Canal in 1929. The Canal is presently operated and maintained by the Army Corps of Engineers, and is still used by pleasure boaters headed to the Albemarle from Chesapeake Bay. Because conditions of the canal may vary, before proceeding boaters should always call the Army Corps of Engineers at the Deep Creek Lock for canal status and lock schedule.

Sites to visit: The site of the Battle at South Mills, with the "roasted" ditch still visible, is about three miles south of South Mills on Highway 343, near the intersection with Sawyers Lane. The canal locks are also still visible at the Canal in South Mills. If you have a boat you can travel up the Feeder Ditch at Arbuckle Landing several miles to reach Lake Drummond—ask about this trip at the Visitor Center. The site of the Half-Way House is on Highway 17 at the Virginia border. The Washington Ditch is accessible on the western edge of the Swamp in Virginia off White Marsh Road (Highway 642).

For more information:
- *Dismal Swamp Canal Visitor Center* (three miles south of the Virginia border), 2356 U.S. Highway 17 North, South Mills, N.C. 27976 (telephone: (919) 771-8333; e-mail address: *dscwelcome@coastalguide.com*

Carolina Journeys

- Alexander Crosby Brown, *The Dismal Swamp Canal* (Norfolk County Historical Society, 1970)
- Bland Simpson, *The Great Dismal* (UNC Press, 1990)
- Thomas Moore, *The Lake of the Dismal Swamp* (1803)(a melancholy poem about searching for a maid who paddles her white canoe at night on Lake Drummond by the light of her firefly lamp).
- For more details on the "roasted ditch" defense and other Civil War skirmishes in this area, read Clint Johnson, *Touring the Carolinas' Civil War Sites* (John F. Blair, 1996).

De Soto Slept Here

Several Spanish expeditions trekked through North and South Carolina in the sixteenth century. The first was led by Hernando de Soto who landed in Florida in 1539 and who proceeded northward through Georgia and into the Carolinas by 1540. I suspect that many of you are like me in that you would like to follow the path taken by de Soto so many hundreds of years ago. But wiser men and women have tried and failed to determine this path. One anthropology professor who has studied the original Spanish texts and compared them with the actual terrain, has concluded that "not a single spot along the route can be ascertained with absolute certainty." Meaning that, there are no "no-kidding-the-Spanish-actually-stood-on-this-exact-spot" places yet established in the Carolinas. Thus, despite all those state historical markers claiming that de Soto passed nearby, it's still just guesswork. But there are tantalizing clues.

Hernando de Soto sailed from Havana to Tampa Bay, in Florida, in May of 1539. His small army consisted of 600 conquistadors, and several tailors, shoemakers, farriers, and trumpeters. He also transported hundreds of pigs, horses and a number of war dogs. De Soto's goal was to find in the southeast the kind of treasure found by Cortez with the Aztecs in 1521, and by Pizarro with the Incas in 1532. De Soto had served as a lieutenant in Pizarro's army and he had seen how the Incan gold and silver had made the invading Spaniards spectacularly wealthy. De Soto intended to deprive the Southeast's indigenous population of the wealth he thought they possessed—by force if force proved necessary. But first he had to find these riches.

De Soto's route in the Carolinas was first officially reconstructed by the De Soto Commission, created in 1935 by the federal government. President Roosevelt appointed a group of scholars to sit on the commission and Smithsonian anthropologist John R. Swanton to lead the group. Studying the original reports of individuals accompanying de Soto, Swanton's group determined that de Soto traveled up the far western edge of South Carolina, entering North Carolina around present-day Highlands, on the Jackson-Macon

County line. Then de Soto turned directly west and passed through the extreme southwestern part of North Carolina—by the present-day towns of Franklin, Andrews, Marble and Murphy—until he entered Tennessee following the banks of the Hiwassee River. This estimated route must have served as the basis for the eight state historical markers in North Carolina that all proclaim the proximity of de Soto's route. These de Soto historical markers are located in Jackson, Macon, Cherokee and Clay counties and each proclaim, maybe erroneously (see discussion below), that de Soto's expedition "passed near here."

Questions and disagreements with the De Soto Commission Report arose soon after its publication in 1939. In the 1980s, Professors Charles M. Hudson and Chester DePratter took a fresh look at the primary source material and the archaeological evidence that had accumulated since Swanton's investigation. Hudson and DePratter decided de Soto's true route lay far to the northeast of what the de Soto Commission had plotted. Their proposed route lay east of Columbia, South Carolina, following the course of the Wateree River, passing near present-day Camden (de Soto's Cofitachequi town) and entering North Carolina near present-day Gastonia. Their route then followed the Catawba River to present day Hickory (maybe de Soto's Guaquili town and Pardo's Guaquiri town), and then turned west past present-day Morganton and Marion (maybe de Soto's Xuala town and Pardo's Joara town). According to Hudson and DePratter, de Soto then marched along an old Indian Trail through Swannanoa Gap and Swannanoa Valley to camp near present-day Asheville—at a Native American village called Tocae. De Soto then followed the French Broad River past present-day Marshall (maybe de Soto's Guasili town and Pardo's Cauchi town) and Hot Springs until the river crossed into Tennessee.

Another writer, claiming to be relying on "the most recent and, to my mind, the most reliable reconstructions of the route," states that from Morganton, de Soto went to the northwest instead of toward Marion. According to this proposal, de Soto followed the route of present-day N.C. 181 to pass by Linville Falls, the Avery County town of Ingalls, and then to the confluence of the North Toe River and the Cane River—where the two converge to form the Nolichucky

River. After spending the night at this location in late May of 1540, de Soto's army then followed the Nolichucky River into Tennessee.

And on the internet you can find even another version of de Soto's route. This one has de Soto passing directly through present-day Columbia, South Carolina, proceeding through present-day Spartanburg, and entering North Carolina near present-day Tryon. According to this account, Tryon is de Soto's Xuala Town. De Soto then went by present-day Hendersonville and then north to Asheville (de Soto's Guaxule town according to this account). The path then led due west, by present-day Fontana Lake and into Tennessee along the banks of the Little Tennessee River.

And, finally, I will not neglect the opinion of my favorite source, the *WPA Guide to the Old North State,* which researched and documented oral histories in North Carolina in the 1930s. The *WPA Guide* describes Skyline Drive which runs along the North Carolina-Tennessee border in the Great Smokies National Park, from Newfound Gap to Clingman's Dome. At the 1.7 mile point from Newfound Gap is Indian Gap—which served as the main gap through the Smokies for an ancient Indian trail. The *WPA Guide* reports: "Tradition is that De Soto and his band crossed Indian Gap in 1540."

So we have at least three different proposed locations for de Soto's passage from South Carolina into North Carolina, and five different proposed locations for de Soto's passage from North Carolina into Tennessee. When the experts disagree so dramatically on de Soto's precise route, what are the earnest avocationalists, like me and you, to do? What should guide us on our weekend jaunts when we seek to stand on the spot where de Soto once stood and survey the scene that he once surveyed? Well ... anecdotes, local legend and tantalizing clues, of course. We must consult the unofficial, unsubstantiated, unapproved body of evidence that waits for us to discover and organize into a coherent whole. Or at least a plausible and enjoyable, albeit not proveable, hypothesis. And such evidence abounds.

Although I've never seen it, there is supposed to be a 450 year old Spanish petroglyph high up on the side of Whiteside Mountain—drive U.S. 64 between Highlands and Cashiers, turn south on Whiteside Mountain Road and proceed one mile to the parking area for the two mile loop trail that goes along the ridge of Whiteside Mountain. It is said that in the 1950s would-be developers of this area discovered a Spanish inscription at the Devil's Courthouse area of the mountain. The letters in the inscription, said to be two inches high and a quarter-inch deep, read: "T.T. Un Luego Santa Ala Memoria." Some say that the carving was made by one of de Soto's men in 1540, and that de Soto's expedition may have followed the old Indian path that ascended Whiteside Mountain from Whiteside Cove on its way westward. A University of North Carolina archaeology professor once told me, somewhat enigmatically, that he had "no direct knowledge" of this inscription but that he "strongly suspect[ed] that it either does not exist or is a hoax."

And there are some old mine shafts—probably originally dug and used by Native Americans both before and after 1540—that contained some artifacts and other evidence that indicated to some that 16th century Spaniards carried out extensive mining at the sites. In the Sink Hole Mine near the little town of Bandana in Mitchell County, and in Tomotla—near Murphy in Cherokee County—mining operations in the mid to late 1800s uncovered remains of old shafts, tunnels and tools believed to have been used by de Soto and his Spaniards when they passed through the region in search of gold. It is said that old pickaxes, cannon barrels, and coin molds were also found. Some of these items may still be in the hands of private collectors.

Amateur archaeologists and private collectors have played a role in the de Soto research over the years. It is reported that in some private collections of artifacts, there are glass trading beads that date from the 16th century—and that these beads were originally gathered from the Peachtree Mound site in Cherokee County near Murphy. This supports Swanton's proposed path, as Swanton suggests that Peachtree Mound is the site of Guasili Town—a Native American village that was described in one of the de Soto chronicles as having such a mound. Then there is the stone with a chiseled Spanish inscription that was found in a farmer's field in central South Carolina—that is now displayed in a local museum. And near a confirmed Native American village site on one of the proposed routes, excavators found a small, decomposed artifact which some have claimed was a piece of chain mail. Surely there is more to be discovered in the accumulated earth and mud of the Carolinas—but such artifacts may never establish with certainty the true paths of the Spaniards.

But, after all this uncertainty and contradiction, I do have my own idea on where de Soto—no-kidding—actually stood at least once in 1540. Mind you, I have no proof—but it seems like a good guess. I have read several reports that the local tradition in Rutherford, Henderson and Buncombe Counties, is that de Soto and his expedition marched past present-day Lake Lure, up Hickory Nut Gorge, past present-day Chimney Rock and Bat Cave, to cross through Hickory Nut Pass, on the way to the Swannanoa Valley and present-day Asheville. An authority no less than the *WPA Guide* (which relied—as I've said—on interviews and oral histories for its information) implies that de Soto passed through present-day Rutherfordton and followed the route of U.S. 74 into Hickory Nut Gorge and ultimately over Hickory Nut Gap. This route was an ancient trail leading up to the plateau (Swannanoa Gap was another) and so is a plausible path for de Soto to have taken. But the most compelling reason to believe that de Soto marched this way is simply the appeal of Hickory Nut Gap itself. It is narrow, well-defined and, except for the asphalt road now topping the gap, it must look much like it did 450 years ago. If de Soto followed the trail up Hickory Nut Gorge then he surely stood in the very spot you are standing—if you have made the trip to Hickory Nut Gap and are standing in the cleft. Beyond a reasonable doubt? Nah. But, you know, it could be ... and it is a great little gap.

Postscript: After leaving the Carolinas, de Soto and his army traveled on through Tennessee, Georgia, Alabama and Mississippi, battling the natives and failing to find the riches they sought. De Soto himself never returned to Cuba or to Spain, instead dying of fever on the banks of the Mississippi River in 1542. The surviving members of his expedition—ultimately numbering only 311—straggled to rescue in Mexico in the summer of 1543. De Soto is often described as the "scourge of the Southeast," for his barbarous treatment of the Native Americans he encountered during his travels. But it is also suggested that greater damage was caused by the Spaniards' introduction of new diseases, like influenza, whooping cough, and smallpox, to the unprotected Native Americans. The archaeological evidence does indicate that the indigenous peoples of the Southeast suffered rapid depopulation in the late 16th and early 17th centuries.

For more on the Spanish expeditions in the Carolinas, see:

- Charles M. Hudson, *The Juan Pardo Expeditions* (Smithsonian Institution, 1990): "We ... discovered why the Soto route had proven so difficult to reconstruct. By themselves, the narratives of the Soto expedition do not contain enough precise information to allow one to plot the route on a map. ... [But Juan] Pardo had visited at least five of the same towns visited by Soto. Obviously, if we could figure out where Pardo went, we should have what no previous Soto scholar possessed—independently located towns in the interior."
- Robin A. Beck, Jr., *From Joara to Chiaha: Spanish Exploration of the Appalachian Summit Area, 1540-1568*, Southeastern Archaeology, 16(2) Winter 1997, pages 162-168: outlines the Nolichucky River route.
- H. Trawick Ward & R. P. Stephen Davis, Jr., *Time Before History: The Archaeology of North Carolina* (University of North Carolina Press, 1999), pages 260-267; see map of proposed routes at page 230. After a review of both the Swanton and Hudson proposed routes and the related archaeological evidence, the authors conclude: "What all this means is that the precise routes of the de Soto and Pardo expeditions through North Carolina may never be established with certainty. The existing documentary records are too vague and will always be subject to different interpretations."
- Timothy Silver, *Mount Mitchell & the Black Mountains* (University of North Carolina Press, 2003), pages 48-55.
- Douglas L. Rights, *The American Indian in North Carolina* (John F. Blair, 1957), pages 6-9.

Meadowmont, Turtle-Handling and the Havacow Hash

As noted above, the Hash House Harrier phenomenon arrived in North Carolina in the early 1980s. Hashers get together on a regular basis to follow an intermittent trail marked with dollops of flour. Hashers are known by their quirky yet somehow appropriate hash nick names, names they earn after proving what they are made of over the course of several hashes. This report, known as the Hash Trash in hashing circles, concerns a late spring hash course laid in Chapel Hill out Highway 54 East by an actor and teacher of actors who once appeared in a TV commercial that also starred a cow—from whence came his hash sobriquet of Havacow. 3Pints is the hash name of an attorney of Irish extraction and SCB stands for a hasher who intentionally short cuts the course—or more precisely SCB stands for "Short Cutting Badperson" or something like that. Meadowmont is now all growed up and we haven't seen the turtle since the incident in question.

Havacow (aka R. Dooley) is a sensitive artist-type sort of guy. He reads Shakespeare. He can talk about incidents which occurred during Oscar Wilde's travels in America. He acts in plays. So what's this guy doing laying a hash if he's such a sensitive artist-type, you ask? I think it is because Shakespeare too would have been a hasher if hashing had been around in those times. Not only are hashes world-round generally populated with more than their fair share of Falstaff and Caliban-types, hash repartee is modeled after the same ribald double-entendre that the bard loved. So I think Havacow agreed to lay this hash because he is gathering material for a screenplay that he is writing and that soon you will see the turtle incident portrayed on the big screen with Leonard DeCaprio playing the 3Pints role and Nicholas Cage as Crowder. Maybe they'll film on location and we can all have bit parts as hashers—although maybe MicroP would be more convincing as the turtle.

Events leading up to and occurring subsequent to the turtle peeing all over 3Pints: It was a hot day for mid-May but ten avid hashers showed up at Havacow's based on the fine reputation of his virginal hash of last year. I personally had some doubts about the hash names that various newbie hashers were seeking to adopt as their own but I figured there was plenty of time to deal with that later—there were bigger problems. For instance early on experienced hashers MiniVann and 3Pints disappeared down a long back check while "Hasn't Paid His Dues" Crowder stood at the check saying that the trail couldn't go that way. Ooooh, that hurt—because Crowder was right. We abandoned Mini and the Pintman and followed . . . ugh, Crowder. Spreadsheet and I opened up, crossed 54 and headed toward the trails on Finley Golf Course—Havacow had done this before. We had a water stop at the bank drive-through opposite Glen Lennox Shopping Center, picked off a few ticks and headed out toward the Friday Center trails. Fairly predictable, 'Cow, but then came the turtle in the road.

It was big, it was bad and it was full of pee—although you couldn't tell just by looking at it. "Pick it up and put it in the woods, 3Pints!" we all screamed. He did, and whether in fright at being so roughly handled by 3Pints or in glee, the turtle turned on the spigot and gave a few pints back to his handler. Big turtles have big bladders, we all noted. 3Pints complained about the smell of turtle pee and no one disputed him. As is often the case in life, there was nothing to do but keep hashing, so we did. Another water break (we all resented this thoughtful treatment and told the 'Cow so) and 'Cow offered us an inglorious short-cut home. Cow-poop, we told him, and Runs With Joints, SlingSlade, Black Shark and Grumpy followed me on to meet our destiny.

Well, Chapel Hill's destiny, anyway. For Chapel Hill had just approved the development of Meadowmont which is the big undeveloped tract of land on the right side of 54 before you get to Finley Golf Course—and Havacow was gonna give us a tour. We ran on old roads between fields and forest, with a few old tumble down houses and barns—lovely, undeveloped country and as far as I know never before hashed. We all picked out the lots we want to buy so that soon we can host Meadowmont hashes all the time—but this was the first Meadowmont hash and we will not soon forget it. Thanks 'Cow. Back at chez 'Cow the SCB's (yeah, I was one, so what?) watched the non-SCB's straggle in—MiniVann and Spreadsheet, Micro, Slingslade, and so on. At the après, we successfully urged 3Pints to leave early ('cause he smelled like turtle pee) and then self-absorbed and full of ourselves we drank cheap beer, talked about Sinatra's acting, Black Shark's grad school shenanigans, what would happen to we fine hashers if Havacow's excitable doggie got let out of the house by mistake, and much more. But you had to have been there. Awn-awn.

A gathering of confused Tar Heel Hashers reviewing the flour trailmarkers

Sunset on Seabrook Island

Printed in the United States
22345LVS00007B/52-60